Raising Wild Ducks in Captivity

Humidity pg. 96

Raising Wild Ducks in Captivity

BY THE

INTERNATIONAL WILD WATERFOWL ASSOCIATION

DAYTON O. HYDE

EDITOR

A Sunrise Book

E. P. DUTTON & CO., INC. · NEW YORK · 1974

FIRST PRINTING

Published simultaneously in Canada
by Clarke, Irwin & Company Limited,
Toronto and Vancouver

Outerbridge & Lazard, a subsidiary of
E. P. Dutton & Co., Inc.

Library of Congress Cataloging in Publication Data

International Wild Waterfowl Association.
Raising wild ducks in captivity.

(A Sunrise book)
Bibliography: p.
1. Ducks. 2. Waterfowl culture. 3. Rare birds.
I. Hyde, Dayton O., 1925– ed. II. Title.
SF510.D8158 1974 636.5'97 73–17075
ISBN 0–87690–111–9

Chapter head illustrations by Cyrille Gentry

Contents

Duck Hunting in North Dakota ED BRY

Editor's Note

From man's earliest moments, he has found waterfowl a delight, an inspiration, and a challenge. Let that reader beware who thinks he can associate with even the lowliest of ducks and escape with his heart untouched.

Some men carve the images of ducks in wood to grace their living rooms or to decoy wild birds in from afar. Others, cramped in blinds, endure freezing discomfort to shoot ducks for the table or to capture their images on film. Schoolboys raise dreaming eyes from books, stare past the Halloween decorations in the classroom windows, to follow a skein of migrating ducks to the horizon. To many, the wedge of wild geese heading north in spring or south in autumn, is the only meaningful harbinger, marking the death of one season in the birth of another.

As a boy I tramped the banks of the Chocolay River in Northern Michigan, not far from where it enters Lake Superior. There to a lone patch of open water flanked by rotting ice, an old squaw, piebald wonder of a duck, bent his whistling wings, filling the astonished air with a whole organ medley of tunes. I dashed home to my bird books, convinced that I had discovered a brand new species. Alas, *Clangula hyemalis* was never destined to become *Clangula hydiae!* Not only had the old squaw duck, dressed in winter elegance, been seen before, but common literature or custom had already given it over fifty different names.

Longing to see ducks closer, I became a hunter, wishing time and again, as my gun cut them down in midflight, that there was a better way. Soon I was only a watcher.

I moved west to Oregon about then, barely in time to catch the tail end of flocks of ducks and geese darkening the sky. Soon these sights were only part of old men's tales, to be scoffed at by the young. And year by year, I became more appalled at man's apathy as all over the land, swamps were being drained, potholes filled. A federal permit to rescue eggs of waterfowl from haystacks and irrigation projects where they were in danger of being destroyed enabled me to raise the young and return them to the wild. Suddenly, I had what I had always wanted: a real, working friendship with waterfowl. From the time I held them newly hatched, wet and helpless in my hands, to the time they circled high above me in the chill autumn migration, I knew them intimately. My joy at seeing their wings beat strong and free as they left me was one no gun could have brought me.

At Yamsi Ranch, where I live in Southern Oregon, some of the lovely clover and bluegrass meadows have become swamps and marshes teeming with insect life. Out of nowhere wildlife arrived to share the ranch with me. Skeins of ducks made racing pylons of the pine trees by the river. Geese hadn't nested there in the memory of man, but I brought in eggs, hatched them, and soon had a migrating flock of several hundred.

I have the life I want now. That wedge of Canada geese, now etched against the sunset, wings sweeping the horizon of the Cascade range in the distance—I am the artist who painted them there if only for this instant. Soon they will swing back over my valley and, clamoring noisy endearments, land at my feet to tug on my frayed trouser cuffs. That cinnamon teal, superbly garbed, that courts his mate there not a dozen feet from me as though I scarce existed—I raised him from an abandoned, field-dropped egg, and now he has brought in a wild female to nest in the tall grasses beneath the canopy of willows.

As dusk comes softly in, I watch as my daughter Marsha feeds a pan of dog meal and milk to a hatching of young ruddy ducks, one of the more difficult species to raise. Without her knowing, the ducks have captured her too, and she will never be the same again.

In one degree or another, the joys of keeping waterfowl may belong to anyone with a bit of water and a patch of land, practically for the asking, and to help others on their way, this book was written. Its contributors are men from varied walks of life, but all have something in common—a fund of knowledge about raising and keeping waterfowl. Most of this has been learned the hard way, by costly trial and error in a field where there were no textbooks. It is their love of waterfowl which has prompted these men to record for you the wealth of their experience, so that you too, professional or amateur, may better enjoy the delights of keeping ducks.

Special credit should go to Dr. William Ferguson, of Winter Park, Florida, who compiled much of the material, and to Marvin Cecil, who built the large Caribbean Garden collection in Naples, Florida, and was the first chairman of the book committee.

A list of the many contributors will be found at the end of the book.

The sponsor of this work is the International Wild Waterfowl Association, a non-profit organization dedicated to the advancement of ornithological and avicultural knowledge of wild waterfowl and to the preservation of rare and endangered species.

Dayton O. Hyde

DR. JEAN DELACOUR

Preface

Waterfowl, that is to say the members of the family Anatidae—swans, geese, and ducks—have probably been kept by men in a state of captivity longer than any other birds. The majority have proven well adapted not only to living but to multiplying under man's supervision, even better perhaps than game birds and pigeons, which, with them, have provided the ancestors of domestic birds.

Greeks, Romans, Chinese, and other representatives of early civilizations first attempted to keep waterfowl as sources of food, and it is not clear when they started maintaining them for the pleasure of watching them. But there were certainly aviaries for that purpose in very ancient times. Like other branches of natural history, bird keeping started in earnest during the Renaissance, and there were important collections of native and even exotic species in Europe in the seventeenth and eighteenth centuries. They grew up tremendously during the nineteenth century when outstanding success was achieved.

Because they are large, conspicuous, and, unfortunately, good to eat, waterfowl and game birds have always been the most persecuted of birds. The pressure on wild populations has been somewhat relieved by domestication and mass breeding of some of them, such as mallard ducks, geese, chickens, turkeys, and guinea hens. But it remains that the wild populations still are decimated for sport or market hunting.

In the course of the last forty or fifty years, a tremendous increase of human power of penetration and destruction has taken place, often obliterating wildlife's indispensable habitats, principally marshes and forests. Very strict measures for preservation of birds and their environment are urgently needed, and their propagation in captivity should be encouraged.

At present, all waterfowl are threatened and unless their protection is insured, they will disappear before very long. But some species are in a more dangerous position than others, particularly those which have a restricted distribution and are adapted to special conditions.

There are now more and better captive duck collections than at any time in history, and more and more, successful techniques for propagation are being developed. Our association has endeavored to gather information from the best known sources in order that many will be able to benefit from the experiences of others.

PART I

The Duck Family

PAUL A. JOHNSGARD

Introduction to the Duck Family

Black-Bellied Tree Ducks

The term "duck" is the collective common designation of most of the smaller species of waterfowl that belong to the bird family Anatidae. This family of nearly 150 species also includes 6 species of swans, 14 species of geese, and 9 species of sheldgeese. Most of the remaining species are considered ducks, in spite of the fact that in a few cases they are called "pygmy geese," "maned geese," or "spur-winged geese."

Since there are about 120 species of waterfowl that can be regarded as ducks, it is necessary that some groupings of related types be used and understood. Although the technical classification of waterfowl has varied greatly over the years, and is still not completely stabilized, some groups are relatively readily designated.

The landmark classification of the family Anatidae that most ornithologists follow, or at least use as a basis for their own groupings, was one proposed by Jean Delacour and Ernst Mayr in 1945. Unlike previous classifications, they utilized a grouping by "tribes," of which they recognized 9. These included the tribes of geese and swans (Anserini), of whistling or tree ducks (Dendrocygnini), of sheldgeese

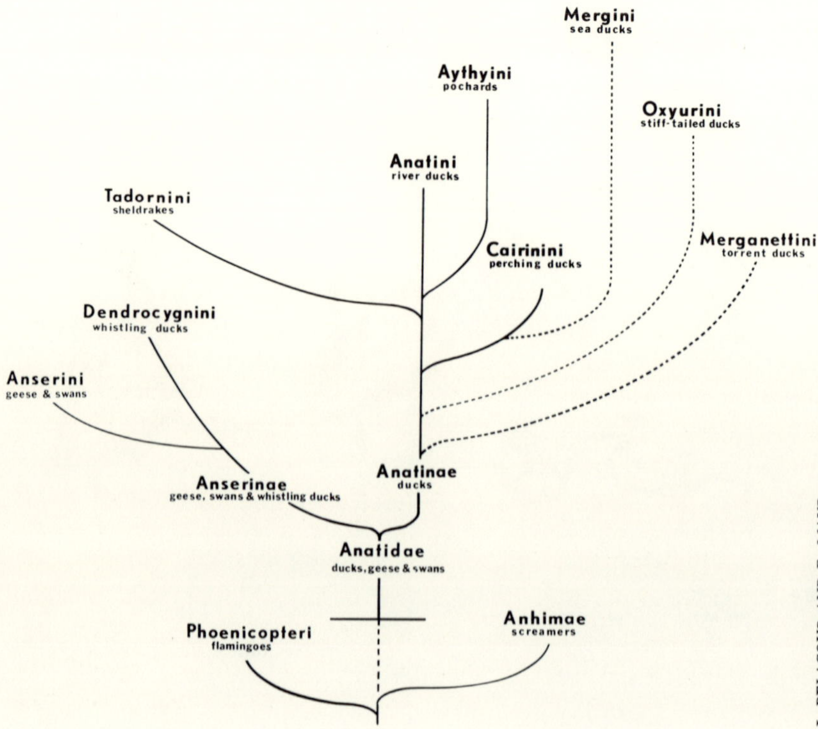

J. DELACOUR AND E. MAYR

Ducks, Geese and Swans. A Phylogenetic Tree of the Anatidae

and shelducks (Tadornini), of river or puddle ducks (Anatini), of pochards (Aythyini), of perching ducks (Cairinini), of sea ducks (Mergini), of stiff-tailed ducks (Oxyurini), and of torrent ducks (Merganettini). More recent research has indicated that some changes in tribal constitution and sequence are probably warranted, but it is convenient here to use this original classification. A diagram of probable tribal relationships, as originally visualized by Delacour and Mayr, is shown in the accompanying illustration.

It may be seen that this present book must deal with a large number of species, some of which have rarely or never been kept in captivity. As an introduction to the various groups of ducks that an aviculturalist might come into contact with, a short description of each of the duck tribes seems warranted.

The tree or whistling ducks are a unique group, since they are actually more closely related to the true geese and swans than they are to other ducks. In these species as in the geese and swans, the sexes are almost impossible to distinguish; both males and females utter whistling notes. All are tropical to subtropical species, and they tend to suffer in very cold weather. They all likewise have strong pair bonds, with both sexes guarding the nest and helping to rear the young. Of the 8 species in the group, the ones most commonly kept in captivity in the United States are the fulvous (*Dendrocygna bicolor*), red-billed or black-bellied (*D. autumnalis*), and White-faced (*D. viduata*).

The shelducks and sheldgeese are in many respects intermediate between geese and ducks, and no single criterion can be used to separate shelducks from sheldgeese other than size. The typical shelducks include seven species (one of which is extinct) in the genus *Tadorna,* many of which are readily adapted to captivity. Although monogamous, shelducks are highly aggressive, and much fighting can be expected if more than one pair is kept on a pond, even if they are of different species. This aggressiveness and tendency toward fighting is offset by the handsome coloring of the birds, which always includes white forewings and a metallic speculum on the wing. Females are usually rather smaller than males and often have more white on the head, especially around the eyes. All of the species are hole-nesters. One unusual South American species, the crested duck, is sometimes considered to be an aberrant shelduck and shares with that group a tendency for fighting and hole-nesting.

The tribe Anatini is the largest of all tribes in the family and includes about thirty species of puddle, dabbling, or river ducks, such as the well-known mallards, pintails, and teal. They are ideal ducks for keeping in captivity, since they mature in one year, have colorful male plumages, have simple dietary requirements, produce large numbers of eggs, and are generally fairly easy to rear. In most species the male is much more brightly colored than the female, and in nearly all species the male abandons the female when she begins incubation and thus normally plays no role in the rearing of the young. Although the males may differ in color among different species, dabbling ducks are all relatively closely related, and hybridization is a common problem under captive conditions. This is especially the case in groups such as the mallard-like species, the pintail group, and the teals. This problem places a limit to the number of species the aviculturalist can safely keep on a pond without running the danger of hybridization.

The pochard tribe Aythyini is composed of about 15 species of

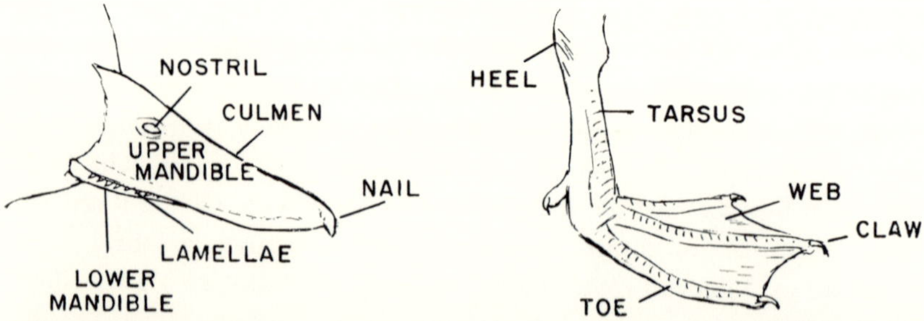

Topographic Anatomy

NEW YORK ZOOLOGICAL SOCIETY

deep-water diving ducks. They differ from the dabbling ducks in their heavier bodies, lack of metallic coloration on the wings, and generally wider and shorter bills. Three species of narrow-billed pochards are somewhat intermediate in anatomy and hybridize fairly frequently with the dabbling ducks. These three species, the red-crested pochard (*Netta rufina*), the southern pochard (*Netta erythropthalma*), and the rosybill (*Netta peposaca*) all breed well in captivity and are desirable species for aviculturalists. The remaining species of pochards include such types as the redhead (*Aythya americana*), canvasback (*A. valisineria*), ring-necked duck (*A. collaris*), and several species of scaups, or bluebills. These birds are somewhat more difficult to keep and breed than are the narrow-billed pochards but nevertheless are all popular with aviculturalists.

The perching duck tribe Cairinini includes a diverse assortment of species ranging from the tiny pygmy geese (*Nettapus*) to the impressively large spur-winged goose (*Plectropterus*). The most familiar species of this group, however, are the Muscovy duck (*Cairina moschata*) and the exotically beautiful mandarin and wood ducks (*Aix galericulata* and *A. sponsa*). These three species are practically the only ones in the entire tribe that the average aviculturalist is likely to consider keeping, although several other interesting species are sometimes kept. Perching ducks are hole-nesters and in nearly all cases exhibit a considerable amount of metallic coloration, especially on the wings. Males are generally much more brightly colored or larger than females and never participate in egg or brood care. The mandarin duck and wood duck are especially notable for their interesting plumage patterns and courtship displays, making them among the most popular of all duck species for aviculturalists.

The sea ducks of the tribe Mergini include a large number of mostly marine or deep-water species that feed to a considerable degree on animal materials. This dependence on a high protein diet, and an unusual sensitivity to heat and such diseases as aspergillosis, makes the sea-duck group one of the most difficult of all waterfowl groups to adapt to captive conditions. Additionally, the birds do not mature before their second year, thus breeding of hand-reared stock requires a minimum investment of two years, and the magnificent male plumage patterns of nearly all the species cannot be enjoyed before that time. Included in the group are the goldeneyes (*Bucephala*), mergansers (*Mergus*), and eiders (*Somateria* and *Polysticta*), representatives of which are all occasionally seen in captivity. Much more rarely seen are the scoters (*Melanitta*), old squaw (*Clangula hyemalis*), and harlequin (*Histrionicus histrionicus*), which provide some of the greatest possible challenges for aviculturalists.

The stiff-tailed duck tribe Oxyurini includes a number of interesting but generally very difficult-to-keep species, of which the best known

Nomenclature NORMA SAFFORD

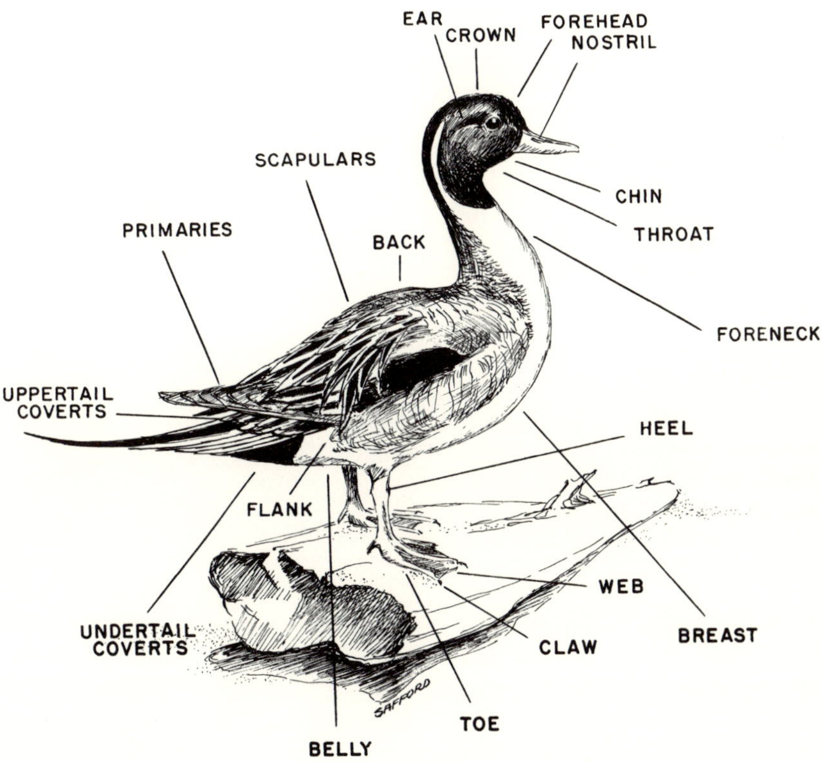

is the North American ruddy duck (*Oxyura jamaicensis*). This small duck is unusually well adapted for diving and as a result is relatively helpless on land. Partly for this reason, stiff-tails are generally regarded as too difficult to warrant keeping in captivity unless they can be provided with natural emergent vegetation in which they can construct their floating nests.

The last tribe recognized by Delacour and Mayr includes only one species, the Andean torrent duck (*Merganetta armata*), which until very recently had never been kept in captivity and might still be regarded as unsuited for consideration as a captive species for all but the most adventurous and optimistic of aviculturalists. The few adults that have been brought into captivity have not survived long, and only very recently have eggs been brought in from the wild and successfully hatched.

As this brief introduction has implied, there are over one hundred species of ducks that have been or are being kept in captivity and most of these have been repeatedly bred successfully under such conditions. The remaining sections of this book should provide the basic information needed by anybody with the inclination to try to keep and breed ducks in captivity, although in the end learning from experience will prove to be the most effective, albeit often the most disappointing, method of learning the principles of duck-keeping. All of the contributors have undergone such experiences, and a careful reading of the following material should help avoid many costly mistakes.

S. DILLON RIPLEY

Ducks on Your Pond*

Wood Ducks

Small beginnings are the best. A very small pond, not too deep, with a shallow place at one end filled with tussocks—that is the most satisfying. Bushes and trees are round about, but not too many. The pond must not be shut in. A grassy bank lies along one side. The birds like to sit out there, preening in the sun, going over all the feathers they can reach with their finetooth comb of a bill, then standing or sitting relaxed, their wings drooping down their sides, primaries half spread. On other occasions they like the shallows and the tussocks. The surface-feeding ducks can dabble there, heads and necks stretching down into the muddy mysteries of the bottom, searching for edibles, insect larvae, small crustaceans, scattered seeds, and fragments of weeds. Sometimes when the water is deep enough they tip up, the forward part of the body stretching after the head and neck, so that only the hinder parts stick straight up out of the water, the feet paddling vigorously to keep them in balance. A duck can maintain this curious pose for a minute or more with ease.

* From *A Paddling of Ducks,* copyright © 1957 by S. Dillon Ripley.

Diving ducks like the deeper water. If the pond is clear and you can see the bottom, it is amusing to throw grain into the water and watch the diving ducks swim eagerly over to the place where the grain has sunk down. They pause a moment, looking about, and then they seem to become smaller, sleeker. It is a muscular effort. The feathers are all drawn down tight, compacted to the body. Then they make a small bobbing jump into the air, not clearing the water. The head and neck are curved downward, and immediately, as they leap up, they start a neat circular course like a half somersault. Down they go under the water with a minimum of splash. In a moment their sleek bodies become visible, outlined by a silvery-gray sheen of air bubbles. Twisting and turning, their wings clasped firmly to their sides, propelled only by their feet, which work together paddling in unison, they seem to dart about the bottom catching up the food grains. Another moment and they simply give up paddling, let the buoyancy of their bodies reassert itself, and rise to the surface again. After three or four such dives they will spend a few minutes rearranging their plumage. This consists of half-standing up in the water and vigorously flapping their wings to shake out any moisture. And then with a deft touch here and there with the bill, they are ready once more.

There is an infinity of motion on a pond. Something is always different, something new. Sleeping or waking, ducks present such a contained appearance, their plumage is so sleek and often so beautifully patterned that they bear looking at endlessly without tiring.

In the early evening the birds are active, swimming about in the dying light. The water is mottled and dappled with broken fragments of color, here pink or orange-yellow, there purple in the shade, there quite white and opaque, reflecting the sky. Constant patterns are being threaded and rethreaded across the surface as the ducks come and go. Each pattern is a series of arrow-like lines stretching away back on either side as that particular duck swims forward. At the point is the bird. Behind, trails a series of constantly widening ripples.

If I am sitting on a bench, perhaps the bench where the feed is stored, the birds congregate toward that side, half curious, half hungry. Some may even begin diving experimentally, hoping that food has already been thrown into the water. One or two walk out onto the bank, and a single bird, tamer than the rest, comes forward and takes a swift peck at those endlessly provocative wormlike objects, shoelaces. The rest stay quiet, loafing. One or two turn their heads to give a quick preen to a single feather on the back or flank. The gesture seems instinctive, without direct purpose, as if to relieve tension. A female wood duck on the bank stretches out her head and neck toward another duck. Her bill opens and shuts a few times in a threatening posture. Only the faintest of noises comes from her opened bill, however. The gesture is more querulous than anything else, and the recipient realizes it and merely shifts its feet slightly without actually moving away.

Now the light is all orange-yellow over the pond, and the shadows round the edges have darkened. The birds gradually move back out into the water without haste. A whisper of sound comes from high over the meadow to the south where the brook runs. The quick whistling wing beats of a flying duck can be heard now, and then a shape is outlined against a brilliant sky. It makes a half-circle and drops in with a quick series of breathtaking sideslips and a final flurry of wings. There is a creaming "plash" noise as the duck glides onto the water with a momentary riffle of spray.

Even in the half-light it is possible to see that it is a male wood duck. His crest erect, he swims rapidly off at a tangent in the van of three or four other male wood ducks on the water near him. All of them erect their crests and compact their feathers so as to show as much as possible in a flank view. Moving almost sideways they steam through the water for a few yards close together. Then the new arrival stops in his course, raises his head high with his crest fully expanded, and makes his sibilant whistle, "peeeet." The noise is rather an anticlimax after such a frenzied performance. But immediately a female answers with her full-throated "oh-eeek" call. This is a penetrating double-syllabled note which always affects the whole population of the pond powerfully. It is evidently a general note of alarm. All the birds react by moving out onto the water and as far away as possible from the silent watcher. The shadows now are so dark that it is impossible to see them any longer. Only the ripples are visible, reflected in the lilac tints from the sky. Faint whispers of duck voices come from the water. The water begins to smoke as the first eddies of mist course across it, ghost-like.

It is hard to move away. There is a spell which lingers over the little pond, compounded of many things, the movement, the color, the voices, the intensity with which the small lives are lived on this blue and green stage. A pond with its busy inhabitants distills a sense of calm and peace amidst the hurry and bustle of life outside. It hums with its own activity, but the effect on the watcher can be that of a cloistered garden, a retreat, a vantage point from which the philosophic-minded, aided by patient observation, can derive much comfort and food for thought.

MICHAEL LUBBOCK AND WINSTON GUEST

110 Dos and Don'ts for Duck Keepers

Canvasbacks

1. Do not keep waterfowl until you have good facilities and sound advice.

2. Do build a vermin-proof fence. An electric wire on the perimeter fence will discourage climbing predators while buried netting placed 2 feet under the soil at a 45-degree angle will prevent predators from digging into the pens.

3. Do provide ample clean water for the number of birds you keep.

4. Do use air compressors to keep ponds ice free in winter.

5. Do provide windbreaks during cold seasons and adequate shade in summer.

6. Do put delicate species in heated indoor quarters during the winter.

7. Do provide ample straw where birds congregate on ice or snow.

8. Do build birds up with extra corn or grain in cold seasons.

9. Do remove all ice and snow from feeding areas.

10. Do not attempt to catch diving ducks on icy ponds.

11. Do ensure that fresh water is flowing through all the ponds in summer to prevent stagnation.

12. Do increase protein in diet one month before breeding season.

13. Do not cut down on protein during the moulting season.

14. Do not feed white or stale bread.

15. Do not neglect feeding grit or oyster shell, especially during the laying season.

16. In humid climates, do not store packaged foods for more than a month.

17. Do remove all stale and wasted feed from the feeding area.

18. Do provide waterproof feed shelters in areas where food remains for any length of time.

19. Do buy your birds from reliable sources.

20. Do not start by keeping delicate species. Suggestions for the amateur collector—decorative, mandarins and wood ducks; dabbling ducks, mallards and pintails; small ducks, Laysan teals and Bahama pintails; shelducks, European; geese, barnacle; swans, black-necked.

21. Do buy in late summer or early fall in order that the birds may become acclimatized and ready to breed the following spring.

22. Do start off with first-year birds. This will give the birds ample time to settle in before breeding.

23. Do not buy so-called "proven breeders."

24. Do quarantine new arrivals for observation.

25. Do double-check your birds before release, to make sure that they are properly pinioned and of the correct sex.

26. Do not place aggressive species with those unable to compete.

27. Do not put more than one pair of a species in a small area.

28. Do not keep closely related species, such as red-billed pintail and Bahama pintail, together or they may hybridize.

29. Do keep extra drakes separate from breeding pairs.

30. Do give the waterfowl time to reach sexual maturity—ducks, from 1 to 2 years; geese, 2 to 3 years; swans, 3 to 4 years.

31. Do provide more than one nest box for those species that require a choice, such as Barrow's goldeneye, chestnut-breasted teal.

32. Do provide various types of nest boxes, such as tree boxes, tunnel boxes, and ground boxes, for the various species, furnishing two boxes for each pair of birds.

33. Do provide teal boxes with 4-inch diameter nesting holes to prevent larger ducks from entering the nesting box and disturbing the eggs.

34. Do not neglect placing peat, damp earth, and dry grass material into the nesting box.

35. Do ensure that all tree- and above-water nesting boxes are provided with ramps leading to the entrance holes.

36. Do make sure that the nesting boxes are still deep enough after nesting material has been added to give the ducks a sense of security.

37. Do give geese, especially the sheldgeese, nesting shelters.

38. Do give swans sufficient nesting material, especially around probable nesting areas.

39. Do provide islands for breeding areas and where possible divide them into separately fenced nesting areas.

40. Do collect eggs daily, preferably in early morning, and replace with wooden dummy eggs. These can be made in the correct size for each species.

41. Do cover the eggs with the surrounding foliage when leaving the nest.

42. Do not take a laying or sitting bird from the nest; wait until she has left to feed.

43. Do not disturb a bird on the nest unnecessarily or she may desert.

44. Do remember that when considerable down appears in the nest this indicates that the bird is ready to incubate.

45. Do not allow small ducks, such as teal, to raise their broods on ponds with other species.

46. Do use an egg candler after a week of setting, to test fertility; discard all clear eggs.

47. Do make a visual check of the eggs daily. If any eggs show irregularity or discoloration, they should be discarded at once, for bad eggs can affect the remainder of the clutch.

48. Do note that if the air sac in the blunt end of the egg increases beyond the norm during incubation a lack of humidity is indicated and the embryo may dry up within the shell.

49. Do store eggs in a cool room. The temperature should not exceed 70 degrees while awaiting setting.

50. Do turn the eggs once daily before incubation and at least twice daily during incubation in machine.

51. Do put eggs either into an incubator or under a bantam foster-mother within ten days of gathering.

52. Do get sound advice regarding operation of your incubator and the proper temperature and humidity for hatching different species.

53. Do disinfect the incubator after each hatch, using a mild non-toxic product.

54. Do remember to clean hands thoroughly before handling the eggs.

55. Do use separate incubators for incubation and hatching.

56. Do use a reliable bantam, such as a silkie-cochin cross. Some breeds are too nervous and make poor foster-mothers.

57. Do give your brood bantam dummy eggs for 2 or 3 days before you entrust her with your valuable hatching eggs.

58. Do not give a small bantam too many eggs to cover.

59. Do not give a small bantam more than 2 goose eggs.

60. Do take note when a bantam dirties her nest. It usually indicates that she is breaking up.

61. Do take note at feeding time. If she is over eager to leave her nest box she may give up setting.

62. Do not use the same bantam for more than 6 weeks.

63. Do ensure that the bantam is free of all vermin.

64. Do make sure the bantam leaves the nest for 20 minutes a day to feed and drink and to cool the eggs.

65. Do turn big eggs by hand.

66. Do keep nesting material in sitting boxes damp.

67. Do not attempt to help the chicks out of their shells.

68. Do pinion the ducklings before they are 5 days old.

69. Do ensure that the ducklings have water within 15 hours of hatching and food within 24 hours. Prior to this time, food is not necessary as the yolk sac has not been fully absorbed.

70. Do start the ducklings on Startina feed and finely diced egg. Feed and change water twice daily and remove any stale food.

71. Do not force-feed young ducklings except as a last resort.

72. Do sprinkle-feed about the brooder and on the backs of ducklings to encourage reluctant feeders.

73. Do place an eager feeder with reluctant feeders of the same size.

74. Do provide live food and duck weed to encourage reluctant feeders.

75. Do use shallow water dishes and grit.

76. Do place small stones in the water dishes to prevent small ducklings from becoming too wet.

77. Do allow plenty of fresh greens for young goslings and cygnets.

78. Do ensure that the brooder is draft-proof and raised off the ground in damp climates.

79. Do not give very young ducklings a large brooder or they may stray from the heat to which they must have access at all times, lest they become chilled.

80. Do provide sacking or indoor-outdoor carpeting inside the brooder.

81. Do not keep young ducklings on wire alone lest they develop callouses.

82. Do clean food and water pans daily.

83. Do clean duckling brooders daily and where possible, move to fresh areas.

84. Do not hose out brooders within the breeder pens.

85. Do provide swimming water for growing ducklings, especially diving ducks.

86. Do not put oily food in duckling swimming water.

87. Do not overcrowd ducklings.

88. Do not mix ducklings of various ages and sizes.

89. Do reduce protein in duckling and gosling diet if wing drooping is prevalent.

90. Do not feed whole corn before ducklings are 6 weeks old.

91. Do gradually replace Startina mash with growing pellets when the ducklings have started to feather up.

92. Do not pinion feathering ducklings lest they bleed to death.

93. Do provide sufficient shade for young ducklings in hot weather.

94. Do isolate broods where there has been mortality until the cause is established.

95. Do not treat sick birds without proper diagnosis. Follow instructions on medication with great care.

96. Do have post mortems done, where possible, on dead ducklings to determine the cause of death.

97. Do take note of any symptoms or irregularities before death which may aid in the diagnosis.

98. Do take extra care when catching young ducklings, since wings and legs are especially fragile.

99. Do not put adolescents and adults together until the young are at least 10 weeks old.

100. Do not attempt to vent sex ducklings without expert advice.

101. Do remember that some species can be sexed by outward

appearance and vocal sounds. Shovelers and pintail females, for instance, have small spots on their bills while the males do not.

102. When shipping birds, do provide crates with ample space.

103. Do provide fresh food and water within the crates.

104. Do make sure that there is ample ventilation within the crates.

105. Do place sacking in the bottom of wooden shipping crates.

106. Do ship early in the week whenever possible.

107. Do give your customer sufficient warning and instructions before shipping your birds.

108. Do label all crates correctly.

109. Do check the waterproofness of new arrivals, especially diving ducks as they may catch cold or drown.

110. Do use wire galvanized after weaving for fencing. Use 1-inch mesh or smaller to exclude vermin and to prevent ducks from inserting their heads in the mesh and injuring themselves.

PART II

Acquiring Stock and Importing Eggs

JUSTIN SOUTHWICK AND PHILIP STANTON

Importation

Widgeons

Let's buy a pair of red-breasted geese from Europe. They only cost \$140 a pair. Here in the U.S. they are worth anywhere from \$200 to \$300 a pair. What a savings that would be! Or would it? The following is an example of costs incurred in the importing of a pair of red-breasted geese:

Cost of red-breasted geese	\$140
Air freight to the U.S.	45
Broker to clear shipment on arrival	35
Bonded trucker to quarantine station	15
Quarantine station fee	15
Bonded trucker to ship back to airport	15
Air express to home	10
Drive to airport to pick up order	
Total Cost	\$275+

One must consider also that the birds could die enroute. This is your loss. Whether the risk is worthwhile or not is a decision left to the individual. Many people do import and find it quite profitable.

Who can import waterfowl? Anyone with the money and the desire to take a chance. The next question one might ask is: how does one go about obtaining birds from a foreign country? For simplicity try the following steps:

I. *Making Application*—Write a letter to the Animal Health Division of the U.S. Department of Agriculture requesting an import permit. After the import permit application is received, submit one copy to the Animal Health Division, Agricultural Research Service, U.S. Department of Agriculture, Federal Center Building, Hyattsville, Maryland 20782. You have to prepare a separate application for each shipment. If your application is approved you will receive three copies of the permit. You then keep one copy of the permit and send the other two copies to the shipper. There is no charge for permits.

II. *The Exporter*—Upon receiving the two permits from the importer, the shipper will keep one for his records and the other will accompany the shipment. He will also have to obtain a health certificate certifying that his shipment is disease free and healthy. The shipper then sends the birds. Waterfowl are usually shipped by air. If the shipment involves larger birds such as swans, the birds are then shipped by boat to minimize shipping costs. Remember, the importer pays *all* shipping costs; this is an important consideration.

III. *Arrival*—Any birds arriving in the U.S. with a value of $250 or more have to be cleared by a broker. This person sees the birds through customs. Orders valued at less than $250 can be cleared by the importer. A person living in Massachusetts, for example, would probably be wise to obtain a broker even on a low-value shipment. The broker's fee would be cheaper than a trip to New York to clear the shipment personally.

IV. *The Quarantine Station*—All waterfowl entering the U.S. have to spend three weeks at a government quarantine station. (Exceptions are made in some cases for Canada.) There are three quarantine centers, located in Miami, Hawaii, and New Jersey. The one most utilized is in Clifton, New Jersey. Here the birds arrive by bonded truck from the airport. They are maintained for three weeks on a standard game-bird diet, as well as fresh "green feed" from a local market. Special care is available at the station at extra cost to the importer. For instance if a certain group of birds require fruit, it is obtained at a local market at retail prices. This could greatly increase the over-all cost. Also, if the birds become ill or act suspicious in any way they are sometimes blood-tested and kept for extended periods of time beyond the required three weeks. This is another extra expense the importer should be prepared to pay.

V. *Effects of Disease and Politics*—Many times when the importer is considering ordering shipments from abroad, he finds certain countries closed or quarantined. For example during 1968, Belgium, Holland, and India were off limits because of the Dutch Duck Plague. Exportation from the British Isles was halted due to serious outbreaks of hoof and mouth disease. For political reasons many Iron Curtain countries are not able to ship livestock. Disease and politics are problems to the importer. A knowledge of those countries free to export wildlife is certainly a basic requirement.

Importation of Eggs

Importing eggs is similar to importing live waterfowl. The aviculturist makes application to the U.S.D.A., Animal Health Center. On approval he receives three copies of the importation permit. He retains one, and two are sent with the order. For example, consider a man living in Worcester, Massachusetts, ordering eggs from Iceland. The price of eggs, depending on the species, varies from $1.50 to $15 each. The beginner might order 30 eggs. The price of the eggs includes the air-express fare to New York. Rather than drive to New York, the breeder will employ a broker to clear his eggs through customs. After the broker clears the eggs he will arrange a flight to Worcester giving the breeder the exact details on their arrival. This is important because the breeder wants to meet the eggs the instant they arrive. The eggs are then picked up and quickly taken to the incubator. No attempt is made to warm the eggs enroute from the airport to the breeder's home. Eggs are packed in plastic containers giving each egg its own compartment. In this situation it costs the breeder $20 for the broker's fees and $7 for air express to Worcester above the initial expense of the eggs.

Purchasing Stock from Local Breeders

Before one can attempt to purchase stock locally he must first check his state regulations. States vary in their laws regarding the keeping of wildlife. It's also a good idea to check local zoning ordinances relative to the keeping of birds.

In Massachusetts, for example, in order to keep wild waterfowl an individual must have a state propagator's permit which reads:

> the licensee above named may possess, propagate, maintain, buy,

> sell, offer for sale and have in possession for the purpose of sale birds and mammals protected by chapter 131, G.L.

This license expires on December 31 of the year issued and costs $3 annually. The stock can be maintained only at the address of the licensee. This permit is given only after a wildlife officer from the Division of Fish and Game has approved the propagating area. A federal permit is necessary if a person intends to sell North American waterfowl. With proper permits people can acquire, propagate, kill for their own use, or release to the wild, captive-reared migratory waterfowl. All waterfowl retained and their offspring must have the hind toe removed from the right foot as a method of identification. All such birds hatched, raised, and retained in captivity shall be so marked prior to reaching four weeks of age.

When a person is ready to acquire stock, it is a good rule to first acquire waterfowl that are not too expensive and difficult to keep. This is the time to make mistakes and learn and profit by them. It is also a good rule to see the birds before you purchase them and try to buy only from established breeders or dealers with a healthy operation. If you buy a pair, make the dealer guarantee, if at all possible, that you are buying a true pair (male and female).

Ordering from a Magazine

Ordering from a magazine is almost like importing birds. The breeder has to figure his transportation costs above the price of the birds. A good example is a man from Maine buying a $3-pair of bobwhite quail raised in Georgia. By the time he receives them, the transportation costs can easily make the total cost of the pair $10 to $12. He might have to pay as high as $7 a pair locally, but it would still be a lot less than ordering a $3-pair from Georgia.

When ordering birds through magazines it is best to make arrangements with the dealer to have birds shipped by air to a large airport. Be sure the dealer knows how to ship the birds properly. Often birds arrive dead or injured due to inadequate crates or crates not able to withstand shipping. Large airports have their cargo offices open 24 hours a day, 7 days a week. Instruct the sender to have the shipment held at the airport and call you regardless of the hour of arrival. If your order is shipped to an airport and then trucked to your local express office, it could often get tied up overnight or by weekends and holidays. This can create dangerous and frustrating situations for both the birds and the receiver. Every system of transportation has its problems. However, air travel and personal pick-up at airports seems best.

Many times the price is not stated in an advertisement. Be sure to write the breeder asking the price per pair and whether or not it is currently in stock. You should also know if you are buying old, young, or proven breeders. Make sure you are not buying a sterile bird, a "troublemaker," or a deformed bird. Most reputable breeders would not offer such birds for sale.

Trading

Trading is also a form of acquiring stock. There are all sorts of trades made between propagators. Many large collections of waterfowl have been obtained by trading. If a beginner starts a collection with some of the inexpensive, easily reared waterfowl, he can often trade several young of certain species for a more valuable pair. This is perhaps the most sensible and rewarding way for persons with limited funds to build up fine collections.

It is always wise to consult local established fanciers for advice when ordering from or trading with unknown sources.

Collecting Stock from the Wild

Federal and state propagating permits *do not* authorize the capture of birds or the taking of eggs from the wild. The U.S. Fish and Wildlife Service will cooperate with scientific agencies, ornithologists, and recognized game breeders and allow them special-purpose permits to collect stock from the wild. Applications for scientific collecting and other special-purpose permits should be made by letter addressed to the regional director at the regional office having administrative jurisdiction over bureau functions in the state where permit activities are proposed. Applications should contain the following information:

1. Name, address, and age of applicant.
2. Species and number of migratory birds or their parts, nest, or eggs proposed to be taken or acquired.
3. Statement of the purpose and a justification for granting such a permit.

GEORGE SEARLES

Importing Iceland Eggs

Ducklings

The importing of eggs from Iceland is a very exciting and rewarding experience. I have been importing eggs for many years, and I am just as excited now when I open the crate as I was with my first order. Many of the ducks that are common now in collections were just brought here by importation of the eggs. Such waterfowl as Barrow's goldeneye were $250 a pair not too many years ago. A pair now can be purchased for about $100. Greater scaup and tufted duck were $75 a pair. They can now be bought for about $25 a pair. Old squaw ducks, red-breasted mergansers, harlequin ducks and black scoters are still very rare in captivity and probably will be so for many years as these species are difficult to raise and maintain in collections.

The first thing to do when you decide to try your hand at raising some of these beautiful ducks is to write the exporter in Iceland for his price list. There is only one person who is engaged in exporting eggs from Iceland—Mr. William F. Palsson, Halldorsstadir Laxardal, S. Ping, Iceland.

Palsson has been exporting eggs for more than 30 years. Eggs vary in price from $2 each for the common species such as European widgeon, tufted, and scaup to $5 for rarer species of ducks such as Barrow's goldeneye and old squaw. Harlequin eggs were $10.50 but the Iceland government has put a ban on the export of harlequin, black scoter, and Northern eider eggs. All of the eggs are collected from the wild and so are almost 100 percent fertile. Mr. Palsson is a very reliable person and will do his best to send you the eggs you order, but, as availability of some species varies from year to year, he may have to substitute to fill your order. You should send in your order and payment for the eggs before the end of March. The eggs are shipped out in the order in which they are received. About the end of April you should write for your permits. You will need one from the U.S. Fish and Wildlife Service. Write to the regional office in your section of the country. Also you will need one from the U.S. Department of Agriculture, Federal Center Building, Hyattsville, Maryland 20782. You will have to write down the number of each species of duck eggs you wish to import. The Fish and Wildlife Service will send you the original and one copy. The Agricultural Department will send you the original and two copies. Send the original from the Fish and Wildlife Service and the original and the yellow copy from the Agricultural Department to Iceland. The other two copies you keep for your files. All checks and permits should be sent by registered mail.

The eggs will usually arrive about the fifth of June, but cold weather and a late spring can delay breeding for a week or more. I have received eggs as late as June 17. All eggs from Iceland enter the U.S. through New York. When they arrive in New York they must be cleared by a customs broker. After they have been cleared for entry into the U.S., they are sent on to their destination. Barnett International Limited, 543 West 43rd St., New York, N.Y. 10035, clears all egg shipments from Iceland. There is a $20 charge by Barnett to clear each egg shipment from Iceland. You should get in touch with them to make arrangements for the payment of the $20 charge and to give Barnett any special instructions you may have concerning the shipment of your eggs. This should be done a month or more before the eggs arrive. Phone number is 212 656-5740.

The man to talk to is Murray Stein. He is a very reliable person and takes great care in handling all egg shipments. He reships the eggs as soon as possible after they arrive in this country, usually within a few hours. When your eggs arrive you can let them set a few hours to settle, or you can put them in the incubator immediately. It really doesn't affect the hatchability of the egg as I have had good hatches either way. The hatch should be from 50 to 80 percent.

MELVIN BLOCK

Care and Handling of Wild Duck Eggs

Baldpate Family

Waterfowl start laying in March and April in most parts of the United States, but the exact time will vary somewhat with the weather.

It is necessary to collect eggs once a day, in the morning after 10:00 or in the evening after 4:30, whichever is more convenient. This will prevent certain predators from stealing the eggs and will slow the habit of egg-eating by some waterfowls.

Ducks will lay their eggs either in open-type or box nests and extreme care should be used in approaching these areas so as not to suddenly scare any female off its nest. This will minimize the chance of breaking eggs should the duck surge suddenly out of its nest.

Eggs are collected and placed in egg cartons or an empty feed bucket. Always place grass or some other soft material in the bottom of the bucket or basket if one is used. In transit from one area to the other, do not over-move, swing, or drop the egg container. Handle this assignment with caution and care.

Thin-shelled, cracked, undersized, or poorly shaped eggs should not be stored for setting. They can be saved and used as feed (boiled eggs) for the ducklings.

While ducks will lay 8 to 30 or more eggs (depending on the type of ducks) in a 10- to 12-week season, it is best to estimate an average of 50 percent good eggs per duck. In the eggs above this number, the life germs are apt to be weak and produce a dead or a very weak duckling.

When gathered, eggs should be placed in a semi-dark, well-ventilated room with a temperature of about 50 to 55 degrees F. The cellar of a house usually provides these conditions best.

The eggs should be turned once each day. The proper method of turning is to incline the eggs 30 degrees to the opposite side of the vertical. The large end of the egg, which contains the air sac, should never be on the bottom.

Duck eggs should not be held in storage longer than two weeks.

PART III

Facilities

CALVIN WILSON

Construction of Pens and Fences

Hooded Merganser

Fencing

At the present time fencing is far more important than in years past due to the increase in hunting dogs. As the population grows in rural and semirural areas more dogs are allowed to roam, particularly at night. It is natural that dogs trained to aid in hunting will do their best to get to confined waterfowl and pheasants.

The best and by far the cheapest exterior protection in terms of years of service is the regular 6-foot, 11-gauge, 2-inch chain fabric. The post supporting this fabric may be wood, but certainly wood is not so durable as pipe.

Many large dogs can easily jump or climb a 6-foot fence, but this can be stopped by using metal arms which support 3 strands of barbed wire. This barbed wire should be of the 4-barb instead of the 2-barb type. This is guaranteed to stop dogs, but does not deny access to cats and raccoons.

Some dogs will dig under the chain fabric to get inside where waterfowl or pheasants are kept. There are two ways to stop this. First if chain link is used, one may purchase used fabric cheaply in 2½-foot pieces and sink in a trench between posts. For neatness this extra fabric may be hog-ringed to the regular fabric. The trench may then be filled in with earth.

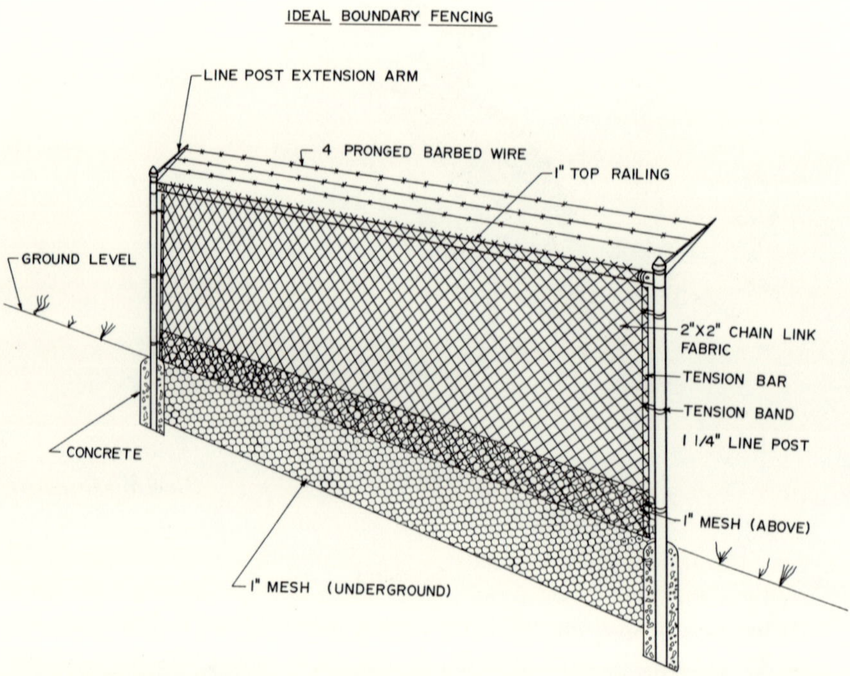

Ideal Boundary Fencing CALVIN WILSON

The second more costly preventative against digging under fencing is to dig a trench some 2½ feet deep between posts and fill this with cement. The width need not be more than 3 or 4 inches.

The first idea mentioned is the least expensive and the material found to be in good condition after 18 years of being covered by earth. It has often stopped small dogs from their attempts to get at both pheasants and waterfowl by digging under the exterior fencing.

The past paragraphs have dealt with keeping out other predators than raccoons, cats, and owls. As chain link is woven in a 2-inch by 2-

inch mesh some thought must be given to keeping in most young ducks as well as small mature birds such as hen bufflehead. Young laysan, falcated, and Hawaiian ducks are prone to wander and of course can be lost. Thus when the outside fence is completed it is wise to install a base wire of 1-inch mesh around the base of the enclosure regardless of how big the installation may be. This idea is very good for most ducks, but young newly hatched teal, mandarin, and wood ducks will go through this 1-inch mesh. To prevent this the waterfowl fancier must use something solid such as 1 by 12 boards at least 2 feet above the water and earth enclosure.

The installation of small-mesh wire fabric may be very expensive but in time it pays for itself in that it does prevent some predators such as rats, weasels, and large snakes from entering the enclosures. Small-mesh chain fabric is indeed costly when installed, but when one takes in account the cost per year of losing valuable species and what one saves in upkeep it is by far the best.

But if the fencing must cross a lake or pond of any kind the plan of using a small-mesh wire is not practical. In this case 1 by 12 redwood boards should be used. Place one above the other to prevent the waterfowl from fence-swimming. This habit not only causes wear on bills and breast feathers but greatly distracts them from nesting. It seems that the habit of waterfowl to head for greener pastures is inherent.

If waterfowl of any kind are kept in pens that are topped with fabric it is wise to pinion them to prevent flying. This does not pertain to wood duck, mandarin, tree ducks, or any of the mallard family. Such ducks are falcated teal, green-wing teal, and any of the more nervous group should be pinioned or at least wing-clipped (one only) to prevent any serious damage and of course in time they will become much tamer.

In conclusion the outside fencing should be the best a fancier can afford. The writer has seen many rare and beautiful ducks, geese, and even swans lost due to poor outside-area fencing. If the fancier keeps other birds, such as pheasants, they too will need the maximum protection along with the waterfowl. Since many waterfowl are prone to wander at night, this attracts the unwanted predators.

Cats and raccoons must of course be trapped. The most effective trap is the National Live Animal Trap. Even if the waterfowl pens are covered with fabric any of the predators can still do much damage by harassment. Great horned owls are very guilty of this. They too may be taken with a large live trap placed in a much used location and properly baited with some form of live food.

Electric fencing is very good in some locations. This requires a location that is strictly rural and where not prohibited by law. A specialist should be consulted before this type of fence is put into use.

Breeding Pens

The size of covered breeding pens naturally depends on the area available and the kind of waterfowl to be kept. For most dabblers and some divers, both North American and imported, the minimum is 6 feet wide and 12 feet deep. This of course is for a single pair of waterfowl. The pond, regardless of structure, should be located at the most distant point from the pen gate as well as away from where visitors may walk. Most waterfowl, unless very docile, are prone to move away from people, and placing the pond location at the most distant point offers them a place to swim. They feel more secure when in water. This kind of pen is certainly not the best for any of the sea ducks or ruddy ducks.

Now if the covered pen is more spacious, say, some 60 feet by 40 feet, then the location of the pond or ponds will be different. Regardless of the kind of pond used, the location should be well away from the

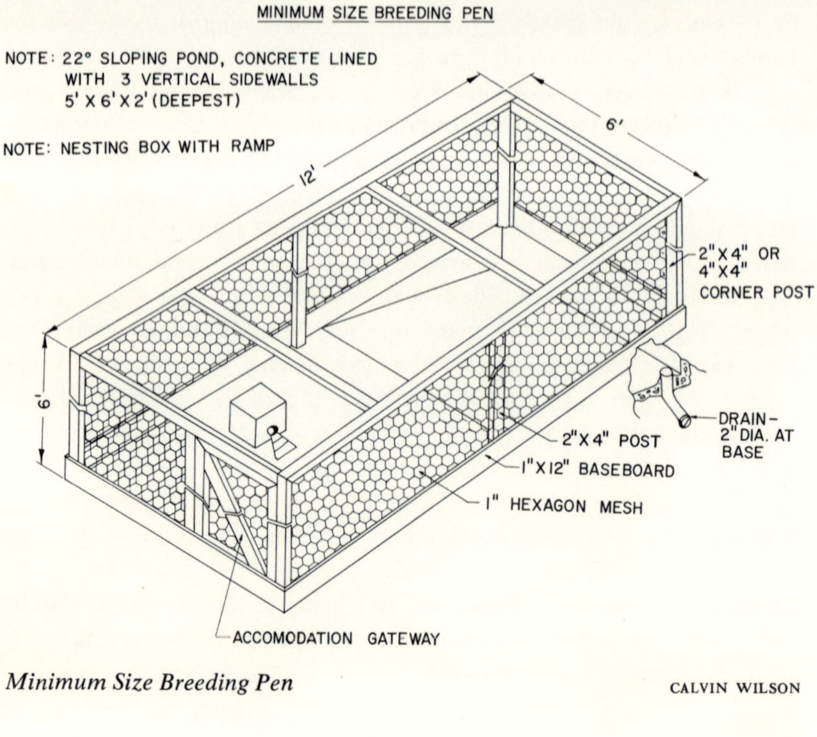

Minimum Size Breeding Pen CALVIN WILSON

outside retaining fence. If cement is used then the ponds should be on ground level for easy draining. In this case all sides should be so sloped for easy access and easy exit.

Thus, in the large area pens, the nest boxes or barrels should be so spaced that there is little territorial conflict or none at all. It is well also to have this large enclosure well planted as many ducks will nest in the grass, and shrubs do offer a good nest site. A pen of this size can take care of 12 pairs of birds unless a pugnacious species is introduced.

JOSEPH J. SCHEDEL
Pens

King Eider

It may be necessary to sink a well to supply water to the duck pond. An artesian well, with its overflow water supply, can be channeled underground to add fresh water to the pond.

Waterfowl property used intelligently is ideal for natural development because it is both enjoyable and interesting. Good judgment and foresight in altering steep slopes or flat areas will enhance its beauty. But erosion and drainage problems must be controlled on such sites. Soil- and water-table conditions should be carefully examined in order to avoid undesirable areas.

Good natural drainage is very important. Surface and subsurface drainage are major problems. Surface water should run off rapidly without soil erosion. Subsurface drainage and control of ground water tables may be very important problems, and in many cases tile must be used.

Proper sanitation requires sanitary sewers to carry off animal waste. It is advantageous to select a site near a natural existing system of adequate capacity—a creek, stream or a river.

Every effort should be made to use open-water areas. Water is an important feature in any waterfowl breeding compound.

All desirable vegetation should be preserved and incorporated into located areas. The planting of large trees is costly, and young trees require many years to reach maturity; therefore, a wooded or partially wooded site is desirable for a natural area for waterfowl propagation.

Almost all waterfowl may be raised successfully in any locality where they can have sufficient water for swimming and a plentiful supply of green forage.

Ducks are rather hardy birds and only require regular shelter from adverse weather. Exotic and rare waterfowls should be confined in a low built house where they will have plenty of room for exercise. Duck houses should be located in an area with sufficient slope to drain it at all times of the year.

The ground for ducks should be sandy soil. The floor in this type of shelter should be dug out to a depth of 6 to 8 inches and be kept continually covered with litter of sawdust, ground corn cobs, dry straw, or any dry absorbent. This litter, whenever it becomes foul or dirty, should be removed and the floor carefully cleaned and replaced with fresh litter.

When possible, it is advisable to include a running stream or small pools of water in each pen. The banks of all pools to which ducklings and goslings have access should gradually slope away from the edge of the pool so that the young fowls will have no difficulty in getting out of the water when tired or when their plumage is watersoaked. No other birds will consume as much water as ducks, and for this reason alone, they must have a plentiful supply of good drinking water constantly. Many young waterfowl have choked to death when feeding due to lack of water or when their feeding troughs are too far away from any source of water.

Water pools will also influence mating of many types of waterfowl.

Compressed air bubblers used in extreme cold winter months will keep water areas open. Both large and small units will do a good job depending on size and depth of ponds.

ELMER HEFT

Ducks in Small Confines with Little Water

Mallard Ducklings

I have often been asked if waterfowl can be raised without a lake, pond, or creek. I have raised most kinds that usually breed near lakes and streams. It *can* be done in a limited space and without natural water. A number of different kinds of ducks can be kept together in the same pen and shelter during the season that they are not nesting. In our northern climate I do this to make the operation more efficient. In the southern states this would not be a factor. In off-breeding season I have about 25 kinds of ducks in a pen 40 by 40 feet. These may include mandarins, wood ducks, European, American, and Chilean widgeon, redheads, canvasbacks, red-crested pochards, rosybills, Philippine, Hawaiian, scaup, ringneck, tufted, Bahama pintails, and tree ducks, which can be kept together in off-breeding season. In our northern climate where they have to be confined in a coop during very cold weather, I find that tufted ducks, scaup, ringnecks, and other less active ducks are better placed in coops alone. Ten to 15 pairs can be kept in a coop 12 feet square if they are allowed to run outside when weather is not too cold. The ducks are fed in the coop and also given water. I use gallon cans cut in two for the water container. If too large a water container is used they will get their bedding wet.

In spring, as soon as our weather warms up, which is several weeks before breeding season, I put pairs into individual pens which are 8 by 18 feet. These pens have been planted to ladino clover and bluegrass the previous fall which gives the ducks nice green feed to eat. I have also planted Japanese yews in most of my pens for shade and for the ducks to hide under if disturbed. I use a nest box 20 inches long, a foot or so wide, and 20 inches high with a hole cut in front about 6 or 8 inches square. Nest box size depends upon the size of the ducks that will use them. I use dry grass for nesting material. For water I make pans about 2 by 4 feet and 6 inches deep from galvanized sheet metal. I make these pans about 3 inches wider on top to keep them from bursting if ice forms in them during the cold weather. For teal and other small ducks a smaller pan for water can be used. I make feeders that are attached to the front of the pen so that feed can be put into them from the outside. For tree ducks, wood ducks, and mandarins I make nest boxes 2 feet deep and a foot square with round holes about 4 inches in diameter. These boxes are nailed to a post on the back of the pen a couple of feet off the ground. If birds are pinioned, a runway must be built up to the hole in the nest box.

During the breeding season I feed a breeding pellet with wheat constituting about one fourth of the total mixture. I keep this feed in the feeders at all times. They also need oyster shells and grit which I mix in with the other feed. After the breeding season I use a holding feed with about one third wheat. I start using the breeding pellets several weeks before the ducks begin to lay.

In certain species of ducks the first clutch of eggs can be taken away and they will lay a second clutch of eggs. I find that this can be done with mandarins, wood ducks, rosybills, pochards, shelducks and tree ducks. Sometimes I leave the young ducklings with the old ones to raise. Other times I take them away and raise them in a brooder.

Young ducks can be fed a duck starter for the first six weeks or so, then duck grower till they are several months old, and then holding feed. Do not feed medicated feed unless it is made especially for ducks.

If one is not interested in raising young, 10 pair or so can be kept in a pen 30 or 40 feet square. This makes a nice display, but breeding success is not as certain as if pairs are kept in individual pens during the breeding season. Pintails, wood ducks, mandarins, widgeons, tree ducks, tufted ducks and teal can be kept in a display pen like this. A galvanized pan 4 by 6 feet can be used for a water container. Or a cement pool can be made for them. The pen can be planted with several Japanese yews or other shrubs. I like the yews as they stay green the year around.

To sum up, those that are interested in keeping ducks can do so without a lot of water or space.

ROBERT V. BAUER

Waterfowl Pen Design

Black Duck

At Pine Forest Farms we wanted a breeding pen that was compact yet provided adequate space for a number of species that had to be penned separately because of their aggressiveness as well as to prevent hybridization. We also wanted an adequate water system to provide ice-free water during the severe winter weather and a good flow of fresh water the year around.

We chose a site in a meadow adjacent to one end of our deep muck-bottomed, spring-fed, 15-acre lake. The land rose gradually from the shore for a distance of about 130 feet.

We put wire over the top of the pen. In order to utilize this wire most efficiently, we made the pen 150 x 144 feet as the top wire used was the 2 x 2⅝ inch rectangular 20-gauge mesh wire, which comes in rolls 150 x 6 feet wide. Within the limits of this pen, we dug two canals at right angles to the lake shore. The canals were 15 feet wide at water level and 5 feet wide at the bottom. The waterway was 3 feet deep and 120 feet long. Sixteen individual pens were set up in this pen, 7 on each of the canals and 2 between the canals with small ponds of their own. Twelve pens on the canals were 60 x 20 feet with two other canal pens

60 x 30 feet. The two center pens were 24 x 75 feet. The pen partitions in the water were made from 1-inch mesh poultry netting the first year. Due to rusting at the water level, this was replaced before the next breeding season by 1 x 4-inch wolmerized treated boards spaced 1 inch apart in a vertical position. We chose the wolmerized treated boards because they do not rot and will not put an oily film on the water as do creosote-treated boards. The perimeter of the pen has a nylon reinforced plastic covering attached to the side wire. This plastic acts as a windbreak and gives the birds more security from dogs and other animals that might threaten the pens. Each individual pen is also isolated from the others with this plastic covering.

For a water supply that would not freeze in the winter, we went to the water pump engineers of the Duro Corporation of Bay City, Michigan. They figured out a formula based on our water capacity and assured us that if we produced a minimum of 60 gallons of water a minute through jets in each of the pens, the water would not freeze over completely at temperatures down to minus 30 degrees F. The smallest pump that they advised us to use was a 7½ h.p. pump with 3-inch suction and 3-inch discharge capacity. The water is pumped in from the lake and is returned to it by gravity flow. The 3-inch suction hose is run out into the lake with the foot valve suspended 4 feet below two 55-gallon floating barrels. A 3-inch header pipe connects to the 2-inch pipes that run along the bottom of each canal. A 1-inch pipe is tapped off each of the 2-inch pipes to supply water to the two small ponds in the center pens. A ¼-inch jet outlet pointing upward is placed at 20-feet intervals in the center of each pen.

This water system has been in use for two years and has supplied us with ice-free water during the winter and a good supply of fresh water the year around. During two weeks of very severe weather when the temperature hovered around the zero mark with a low of minus 21 degrees F. and a high of plus 10 degrees F., some ice formed along the shoreline and along the board pen partitions.

We do have erosion problems. Our soil is of a sandy, gravel, and muck variety which crumbles readily in water. The use of log and plank shoring up to the water level has only proved partially successful. A low profile grassy bank above and at water level might solve this problem. We lost a few birds that caught their heads in between the partition boards in the water.

The basic water system can be used in a variety of ways to provide ice-free, good quality water the year around. The water can be pumped from a stream through a series of small ponds and back to the stream. The 2-inch pipe can be laid out in grid fashion on a single large pond or lake to provide ice-free conditions in the winter and a good quality of water the year around. Even those without a water supply but adequate drainage facilities could put in a 2 to 4-inch well depending on size and number of pens contemplated.

Our second pen of the two-canal design does not have partitions in the one canal, so it provides a flowing pond 120 feet long by 15 feet wide, and is used by a variety of birds. We have had success in getting over 37 species of waterfowl to produce for us with this system.

CALVIN WILSON

Artificial and Natural Ponds

Mallards

An important necessity for small 6 by 12 pens is the cement pool. As stated in a prior chapter the pool should be located away from any disturbance. Three sides of the pool may be perpendicular but the side where the waterfowl enter should be sloped at about a 22-degree angle and the cement well serrated. It is always well to have a large valve to permit a rapid drain to take out as much of the accumulated mud as possible.

If no area or drainage system is provided to accept the pool water when a change is needed, the pool will have to be built above the earth area so the water will drain onto the surrounding earth. No cement pond that cannot be drained has any merit as all waterfowl are prone to carry into the pond a large amount of dirt. This of course just becomes mud in the pond. Also in a short time the fowl will contaminate the water with their own wastes.

The maximum depth of a cement pond need not be very great. A good depth for average waterfowl use is 12 inches or perhaps just a little more. This is not the best for the ruddy duck group but would do if they were not kept in this condition too long.

In regard to the water needs, it is best to consider the small enclosure first. In this case let us assume that the fancier has only the water supply from the house water pipe line. This is fine for the small cement pond as a small amount can be regulated and left to drip at all times. This means that the pond need only be drained once a week until the hen starts to nest. The fancier must use his own judgment on how often to drain. This of course depends on how messy the hen duck is by nature.

If a series of breeding pens are built over a channel of moving water there is really no problem in regard to draining. Boards should divide the pens so the birds will be more content and not swim back and forth to get into the adjoining enclosures.

The large ponds are generally natural and in many cases are fed by either springs or a stream. If the water supply is ample then there will be no stagnant conditions. The only worry of course is that the intake and the outlet are so arranged that the waterfowl do not escape. Flood conditions must be thought of and steps taken to prevent washouts. A well-worked-out spillway will prove very effective.

The fortunate fancier with the natural lake regardless of the source of water should at all times have waterfowl traps so located that any of the young or old can be trapped either for shipping or pinioning. It is assumed that the lake will not be enclosed on top and thus trapping is necessary. By careful feeding most waterfowl can be trapped.

GEORGE W. MILLER

Bubblers

Redheads

A method to keep ponds from freezing over in winter and to provide open water for waterfowl is in use at "Fairlake," R.R. #3, Ayr, Ontario, Canada.

In the fall of 1964 we heard from a friend that he was keeping his trout pond from freezing over by using two small diaphragm-type air compressors. We decided to make a test run on our lake, in water about 8 feet deep, to see if the compressors would keep our waterfowl ponds open for us. We bought two air compressors and had them mounted on a piece of 6-inch channel iron with a third horsepower electric motor mounted between them. They were each connected to half-inch plastic pipes running out into the lake. Both pipes were plugged at the other end and 14 holes were drilled 8 feet apart starting at the plugged ends using a No. 64 drill. The pipes were weighted with ⅓-inch iron rods fastened to the pipes with wire ties every 4 feet. Reinforcing rod is best and cheapest. We kept an area of about 40 by 120 feet open all winter with very little trouble and replaced only one diaphragm all winter. The ice on the rest of the lake froze to about 16 inches in thickness.

Knowing it would work, we fenced in part of the lake to be used as a breeding area for diving ducks and coots. We have an area of 70 by

144 feet of the lake enclosed; the deepest part is about 8 feet. We are using three compressors to keep it open, so far with good results. On one rare occasion the mercury dropped to 17 degrees below zero with a high wind, and the birds still had plenty of open water.

We are using one compressor to help the springs in one of the ponds with fair results. Here the water is only about 3 feet deep. Another one is being used to keep open water in a fairly large shallow pond where we winter our smaller kinds of geese. In this case we found that we had to put the air holes closer together, about 3 or 4 feet apart, to be of any use. Even then we only had small open holes in severe weather but the geese did have some open water to swim in and to drink.

Some conclusions: We now know that a small diaphragm type air compressor with air holes in the pipe every 8 feet (not more than 15 holes, using a No. 64 drill) will keep an area of about 20 by 120 feet open, somewhat less in severe weather, in water 6 feet or more deep. In shallower water the air holes will have to be closer together—about the same distance apart as the water is deep. The air holes could be a bit

Air Compressors Showing Hook-ups GEORGE MILLER

larger, but naturally you will not be able to keep as large an area open.

It is advisable to have at least two compressors in use at all times; in case one breaks down (and they do) you will still have open water. From what we now know, single unit compressors powered with quarter horsepower, continuous run, electric motors are advisable. A good idea is to have a spare unit on hand; in case of a breakdown disconnect the unit and connect up the spare. Remember the compressors *must* run all the time from freeze-up until spring. Occasionally an air hole in the pipe will plug up and the pipe will have to be lifted, the hole cleaned out, or a new one drilled.

We do not think they will be of any use in very small ponds where the water is less than 3 feet deep. There is simply not enough latent heat on the bottom of the pond. However if your pond has a fairly strong spring, a compressor would be a help.

By using bubblers we are able to winter swans, geese, and ducks (other than the very tender kinds such as the whistling or tree ducks)

A Single Unit Hook-up GEORGE MILLER

outside all winter. The only shelter they have is a few sheets of plywood and some evergreens on the north and west sides. Among others Australian and European shelducks, Chiloe widgeon, Bahama pintails, cinnamon and green-wing teal, wood ducks and mandarins are wintered this way with very few losses. The plumage is kept in better shape also than if they are wintered in a building.

WILLIAM CONWAY

The Value of Plant Cover in Duck Enclosures

Mandarin Drake

The development of many kinds of duck-nesting boxes, shelters, islands, or baskets is an attempt to offer captive waterfowl a substitute for the kind of plant and natural cover that they might find in the wild. This attempt has proven successful with many species; nevertheless the development of growing plant cover around waterfowl ponds, especially duck ponds, has great value in captive waterfowl management.

Individuals of many species are thought to breed more readily when a nesting-shelter opportunity is comparable to those they have used before, or in which they hatched, and some birds seem reluctant to use artificial shelters. The addition of "natural" plant cover to the banks and surroundings of waterfowl enclosures increases nesting opportunities, enlarges the effective space of the enclosure by breaking up its area, and allows less dominant birds to escape the surveillance of more aggressive individuals. Natural cover also improves conditions for over-wintering by forming windbreaks. Finally, good plantings make the waterfowl pond look better.

Of course, collections in various parts of the United States have different planting possibilities. High clump grasses (difficult to maintain in goose collections) and low shrubs are selections offering considerable environmental variety to the enclosure. Grasses successful at the New York Zoological Park include hardy forms of *Eulalia* or *Miscanthus* as well as *Erianthus*. Popularly called "Eulalia," "fountain grass," or "hardy pampas grass," various kinds are commonly available in nurseries. True pampas grass (*Cortaderia*) is not hardy over most of the northern United States. Various hardy bamboos (consult your local nursery) also provide excellent cover, and water plants such as phragmites and cattails are attractive and useful where conditions for their specialized culture are present. Low-growing junipers and even yews have proven successful in many collections.

In non-breeding seasons, enclosures may be managed to protect their plantings by applying the principles of crop rotation. Areas showing wear are secured from the depredations of the birds with temporary fences and given time to recover.

Unfortunately, more plantings mean that more vigilance in policing and sanitation must be exercised. It is obviously more difficult to police a planted enclosure for decaying materials or a sick bird than an open one.

CALVIN WILSON

Nests

Buffle-Heads

Even though an enclosure may have vegetation, it is wise to also provide nest boxes which suit the species kept. For the diminutive teal, and up in size to pintails and redheads, the nest box should be at least 2 feet long, 16 inches wide and 16 inches high. The box should have a round entrance at each end so that the hen can escape predators, and the entrance should be close to the bottom of the box for easy access. In most cases, it is good to leave the box without a floor so that the nest may be made in a hollow in the earth and lined with hay or weeds.

In the case of mandarins, wood ducks, hooded mergansers, and any of the tree ducks, the nest box should be raised off the ground, with a ramp leading to a landing platform at the nest box entrance. The ramp should not be too steep and should have plenty of cross-pieces to help the ducks climb.

In small pens, the box should not be placed near the entrance gate but should be well shaded and hidden in vegetation.

Some propagators who have several species breeding on sizable ponds place nest boxes out over the water on stilts or floats. Here again care must be taken that the birds are able to enter the boxes.

ED BRY

Wood Duck on Nest. Hardware Cloth Aids Young in Climbing Out of the Nest

ROBERT V. BAUER

A Type of Nest Box Bauer Found Successful for Red-Crested Pochards, Redheads, Cape Shelduck, Ruddy Shelduck, Ruddy Ducks, Philippine Ducks, Mexican Ducks, Pintail, and Bahama Pintail

Nest Boxes

DONALD P. WELLES, JR.

ROBERT V. BAUER

LEFT: *A Box Used by Wood Ducks, Mandarins, Chestnut-Breasted Teal, and Mexican Ducks*

RIGHT: *Nest Box Interior*

ROBERT V. BAUER

ROBERT V. BAUER

A Box Used by Canvasbacks, Redheads, Greater Scaup, Red-Crested Pochards, Mexican Ducks, Rosybills, Indian Spotbills, Cuban Tree Ducks, Blue-Winged Geese, and Barnacle Geese

PART IV

Breeding

G. MICHAEL FLIEG

Breeding Waterfowl

Tufted Ducks

Fertility in egg production in waterfowl depends on three factors, the first and foremost being nutrition. The factors entering into nutrition are many. For instance, in egg production, vitamin B6, Biotein, and Choline (which is adequate in most prepared chows), adequate calcium, and vitamin D3 are important for building the egg shell, while vitamin B2 is vital for maximum egg production. For maximum hatchability, vitamins E, B12, D3, and B2 and manganese are very important. Also necessary are Pantothenic acid (source, trout chow), vitamin B6 (adequate in most prepared chows), and Biotein (found in fresh greens). Fitting up a proper diet in regard to these vitamins isn't really too difficult. Vitamins E and B2 are found in an egg base supplement called LIV.[1] Vitamin B12 and manganese are readily available in Purina Trout Chow. Free choice oyster shells supplement the extra calcium requirement of the female during laying, and vitamin D3 can be obtained by sunlight, but must be supplemented if the birds are kept indoors. It is also essential that fat be kept at a minimum in the diet. It is a known fact that fat birds, espe-

[1] LIV—available from Farm & Wildlife Products Inc., Omaha, Nebraska 68106.

cially males, cannot produce fertile sperm if the testes are too well insulated with a hot layer of fat.

I recommend seasonal feeding of waterfowl, therefore, as follows:

Spring & Summer—	½ Purina Trout Chow
	½ Purina Game Bird F&M Chow Supplemented with LIV, oyster shell, and greens
Fall —	¾ Game Bird Chow—Purina
	¼ Trout Chow—Purina
Winter —	½ Purina Game Bird Chow
	¼ Purina Trout Chow
	¼ Scratch[2]

The next factor could probably be called duck psychology, as breeding birds must be physically and emotionally ready to breed. Proper nesting sites are, therefore, essential. Well planted island and shore lines for good cover are needed so that the birds can escape other birds and man for nesting duties.

Moving birds to different areas should be avoided during the nesting season, although new surroundings have been known to stimulate breeding. Some birds should be penned individually for breeding purposes, especially the more dominant and most docile. Even on large bodies of water, a few dominant pairs can disturb the nesting of most of the other birds. Nesting areas should be fenced off from one another and be in abundance if all of the species are expected to nest. Of course, clean nesting material is essential for good eggs so they will not become too wet or too dirty on contact with the ground.

Unbalanced sex ratio should be avoided, if possible, for an abundance of drakes could cause undue stress, not only on the ducks of their own species but also on those of closely allied species they would hybridize with. Too few drakes, on the other hand, may not be able to properly fertilize all of the eggs from all of the ducks; thus resulting in lower fertility. It is always well to hang onto a pair of offspring each season, as older birds soon lose virility and their production decreases after a few peak years. Some birds are too young to nest even when they reach the age of sexual maturity. Birds which are imprinted on man or other species should be separated from their breeding flock to avoid hybridization.[3] Do not expect Southern hemisphere birds, such as Australian shelducks, which reverse the seasons, to start nesting immedi-

---

[2] All Purina Chows—Ralston Purina Company, St. Louis, Missouri.

[3] Remember that captive-reared birds will most often nest profusely, while wild caught birds rarely ever nest, at least for quite a few seasons; so if possible for your breeding stock, get captive-reared birds.

ately, even though they reach the age of sexual maturity. Sometimes it takes them 7 to 8 years so that they are reversed for nesting during our spring and summer.

The first eggs of the season, as well as the last eggs, tend to be infertile, as well as eggs layed and left untended during inclement weather.

Care of eggs is a third very important factor in ensuring good production and fertility. First and foremost in this area is the abundance of clean nesting material for the birds.

Eggs should be washed in a good commercial egg cleaner. If you decide to store eggs or clutches, store them at 40–55 degrees but not over 7 days. Fumigate your incubator periodically. Candle the eggs properly so that infertile eggs or bad eggs will not stay in the incubator to explode and encourage other problems, such as bacterial infections with the other eggs.

DONALD BRUNING

A Summary of Incubation Periods

A Feeding Mallard

EDITOR'S NOTE. For a particular species, incubation periods often vary as to the authority. Although the periods listed here are useful guidelines, the individual operator will do well to work out his own summary.

While the names and systemic list of this summary of incubation periods were taken from Delacour, the periods came from a variety of sources, including Frith's *Waterfowl in Australia;* records of William Macy, Long Island, New York; Gus Griswold of the Philadelphia Zoological Park; M. W. Weller; Delacour; the Cornell Laboratory of Ornithology; New York Zoological Park; and Michael R. Lubbock of the Wildfowl Trust in England.

Common Name	*Scientific Name*	*Incubation*
Spotted Whistling Duck	Dendrocygna guttata	31 days
Plumed Whistling Duck	Dendrocygna eytoni	28–30 days
Wandering Whistling Duck	Dendrocygna arcuata	about 30 days
Fulvous Whistling Duck	Dendrocygna bicolor	28–32 days

Common Name	*Scientific Name*	*Incubation*
Black-billed Whistling Duck	Dendrocygna arborea	30 days
Indian Whistling Duck	Dendrocygna javanica	probably 28–30 days
White-faced Whistling Duck	Dendrocygna viduata	28–30 days
Red-billed Whistling Duck	Dendrocygna autumnalis	27–28 days
Australian Shelduck	Tadornia tadornoides	30 days
Paradise Shelduck	Tadornia variegata	30 days
South African Shelduck	Tadornia cana	30 days
Ruddy Shelduck	Tadornia ferruginea	28–30 days
Radjah Shelduck	Tadornia radjah	30 days
Common Shelduck	Tadornia tadornia	30 days
Crested Duck	Lophonetta specularioides	30 days
Magellanic Flightless Steamer Duck	Tachyeres pteneres	probably 31 days
Falkland Flightless Steamer Duck	Tachyeres brachypterus	probably 31 days
Flying Steamer Duck	Tachyeres patachonicus	unknown
African Black Duck	Anas sparsa	28 days
Bronzed-winged Duck	Anas specularis	29–30 days
Northern Mallard	Anas platyrhynchos	26 to 28 days
American Black Duck	Anas rubripes	26 to 28 days
Spot-billed Duck	Anas poecilorhyncha	26–28 days
Grey Duck	Anas superciliosa	26–28 days
Philippine Duck	Anas luzonica	25–26 days
Meller's Duck	Anas melleri	28 days
Yellow-billed Duck	Anas undulata	26–27 days
Grey Teal	Anas gibbernifrons	24–26 days
Madagascan Teal	Anas bernieri	
Chestnut Teal	Anas castanea	23–26 days
Brown Teal	Anas aucklandica	28 days
South American Green-winged Teal	Anas flavirostris	24–26 days
Northern Green-winged Teal	Anas crecca	21–24 days
Baikal Teal	Anas formosa	25 days
Falcated Duck	Anas falcata	24–26 days
Gadwall	Anas strepera	25–26 days
European Wigeon	Anas penelope	24–25 days
American Wigeon	Anas americana	24–25 days
Chiloe Wigeon	Anas sibilatrix	24–26 days
Bahama Pintail	Anas bahamensis	25–26 days
Chilean Pintail	Anas spinicanda	25 days
Hawaiian Duck	Anas wyvilliana	26 days
Laysan Teal	Anas laysanensis	26 days
Blue-billed Pintail	Anas acuta	26 days
Red-billed Teal	Anas erythrorhyncha	25 days
Marbled Teal	Anas angustirostris	25 days

Common Name	*Scientific Name*	*Incubation*
Cape Teal	Anas capensis	25 days
Hottentot Teal	Anas punctata	24 days
Silver Teal	Anas versicolor	25 days
Garganey	Anas querquedula	23 days
Blue-winged Teal	Anas discors	24 days
Cinnamon Teal	Anas cyanoptera	24 days
Red Shoveler	Anas platalea	25 days
Cape Shoveler	Anas smithi	20 days
Australasian Shoveler	Anas rhynchotis	26 days
Northern Shoveler	Anas clypeata	26 days
Salvadori's Teal	Anas waigiuensis	unknown
Ringed Teal	Anas leucophrys	23 days
Indian Pink-headed Duck	Rhodonessa caryophyllacea	unknown
Australian Pink-eared Duck	Malacorhynchus membranaceus	26 days
Mountain Blue Duck	Hymenolaimus malacorhynchos	30–35 days
Andean Torrent Duck	Merganetta armata	unknown, believed over 38 days
Australian Freckled Duck	Stictonetta naevosa	26–28 days
Steller's Eider	Polysticta stelleri	
Common Eider	Somateria mollissima	24 days
Spectacled Eider	Somateria fischeri	24 days
King Eider	Somateria spectabilis	22 days
Red-crested Pochard	Netta rufina	26–28 days
Rosy-billed Pochard	Netta peposaca	26–28 days
Southern Pochard	Netta erythrophthalma	26 days
Canvas-back	Aythya valisineria	26 days
European Pochard	Aythya ferina	27 days
Redhead	Aythya americana	28 days
Madagascan White-eye	Aythya innotata	26–28 days
Australasian White-eye	Aythya australis	26 days
Ferruginous White-eye	Aythya nyroca	26 days
Baer's Pochard	Aythya baeri	27 days
Ring-necked Duck	Aythya collaris	26–27 days
Tufted Duck	Aythya fuligula	25 days
New Zealand Scaup	Aythya novae-seelandiae	26 days
Lesser Scaup	Aythya affinis	27 days
Greater Scaup	Aythya marila	27 days
Brazilian Teal	Amazonetta brasiliensis	25 days
Maned Wood Duck	Chenonetta jubata	30 days
Carolina Wood Duck	Aix sponsa	28–32 days
Mandarin Duck	Aix galericulata	28–30 days
Comb Duck	Sarkidiornis melanotos	30 days
Muscovy Duck	Cairina moschata	35 days
White-winged Wood Duck	Cairina scutulata	30 days

Common Name	*Scientific Name*	*Incubation*
Hartlaub's Duck	Cairina hartlaubi	32 days
White-winged Scoter	Melanitta fusca	27–28 days
Surf Scoter	Melanitta perspicillata	28 days
Black Scoter	Melanitta nigra	27–28 days
Harlequin Duck	Histrionicus histrionicus	30–32 days
Long-tailed Duck	Clangula hyemalis	23–24 days
Common Goldeneye	Bucephala clangula	26–28 days
Barrow's Goldeneye	Bucephala islandica	30 days
Buffle-head	Bucephala albeola	22 days
Smew	Mergus albellus	28–30 days
Hooded Merganser	Mergus cucullatus	28–31 days
Brazilian Merganser	Mergus octosetaceus	unknown
Auckland Islands Merganser	Mergus australis	unknown
Red-breasted Merganser	Mergus serrator	26–30 days
Chinese Merganser	Mergus squamatus	unknown
Goosander	Mergus merganser	28–32 days
Ruddy Duck	Oxyura jamaicensis	20–24 days
Argentine Ruddy Duck	Oxyura vittata	21–25 days
Maccoa Duck	Oxyura maccoa	24 days
Blue-billed Duck	Oxyura australis	26–28 days
White-headed Duck	Oxyura leucocephala	unknown
Masked Duck	Oxyura dominica	unknown
White-backed Duck	Thalassornis leuconotus	26–31 days
Australian Musk Duck	Biziura lobata	unknown
Black-headed Duck	Heteronetta atricapilla	24–25 days

GEORGE SEARLES
Incubation by the Female Duck

Pintail Drake

In discussing the natural incubation by a female duck, we must remember that really it is natural incubation in an unnatural environment. The bird has been taken from its natural habitat and placed in a man-made breeding pen. There are many advantages and disadvantages to letting a duck incubate eggs. It is easier and less work to incubate eggs under a duck, as you are relieved of the turning of the eggs. Also the duck can better adjust the temperature and humidity than is possible in an incubator. If the eggs are left with the mother to hatch she will rear them without too much difficulty. The ducklings and the mother should be in a pen by themselves. If you keep them in a mixed collection of ducks, the other ducks will kill the young. The exception to this rule is the eider ducks. Sometimes in the wild two or three mother ducks have been seen with 30 to 50 young.

The mother should be bothered as little as possible when she is incubating. Some ducks will let you take them off the nest, check the eggs, and replace them on the nest, seemingly indifferent to what is happening. Muscovies are very good mothers, as are mallards. I once had a wild mallard nest in one of my ponds. I replaced her eggs with some old squaw eggs. Whenever I went to check the eggs she would attack me and get hold of my pants cuff. I could pick her up in my hands and then she would bite me. I dread to think what would happen to such a devoted mother if she were visited by a fox or mink while nesting.

Small eggs should never be put under a large duck. Also all eggs should be about the same size. A duck will sometimes desert her nest if the eggs are of different sizes. She may also roll the odd-size eggs out of the nest.

A duck should always be left to find her own nesting site. Once she has decided when she wants to nest, she should not be moved. Very few ducks will continue setting if the eggs are moved to a different location. The duck should be allowed to set a few days on dummy or infertile eggs to make sure she is ready to incubate. After she has been setting a few days the eggs can be replaced with whatever eggs you have decided to incubate under her. The duck should never be disturbed while she is incubating. The eggs should be switched when she leaves the nest to feed. The eggs should also be candled when she leaves the nest; with a little watching, you can find out her schedule. She will leave about the same time each day. A muscovy will continue to set for up to 60 days, while most mallards will leave the nest within 35 days. Near the end of the incubation period she will set more closely and will leave the nest less often and for shorter periods of time. It is only normal to want to see the ducklings as they are hatching. This is very dangerous as the mother may get very nervous and step on the ducklings and kill them. The ducks may be left with the mother or taken away and raised in a brooder. In a brooder you have more control over the young. I prefer to raise the young in brooders, except for swans which are better left with the parents.

There are other disadvantages to using ducks for incubating eggs. If the ducks are kept in a mixed collection it is almost impossible to raise young. For, as was stated earlier, ducks have a tendency to kill each other's young. If you let the female duck set on her own eggs in a mixed collection you can lose a lot of eggs. I once let a female rosybill set on 8 of her own eggs. When I went to candle them 10 days later there were 19 eggs in the nest. Other ducks had been using the nest as a dump nest. Needless to say, all the eggs were spoiled. There were, in addition to the 8 rosybill eggs, four pintail, three redhead, two chiloe widgeon, one European shelduck, and one greater scaup. I find it is best to collect the eggs each day, leaving dummy or infertile eggs in each nest so the female will not desert the nest. Sometimes you will still find the eggs of two or three different species in one nest. I think the best way to incubate duck eggs is to let a duck or hen set on the eggs for the first week or 10 days. Then take the eggs away and put them in an incubator for the rest of the incubating period. You can then put a new setting of eggs under the female. This way you can actually set 36 to 48 eggs under one hen or duck over a period of 28–30 days. Using this method I get an 80 percent to 90 percent hatch. I know of one man who will not use incubators and uses only muscovy ducks and bantam hens to hatch eggs. Everyone will have to find out what works best for him, as one man's meat is another man's poison.

ROBERT V. BAUER

Natural Incubation of Ducks

Pintail Family

Mallards

In the areas of the country where I have observed nesting waterfowl—New York and lower Michigan—the mallard nesting season begins about the end of March and continues through June with some birds occasionally nesting into August. The average clutch size of birds that have been in captivity for many generations is 12 eggs (if left for natural incubation); if picked up, the average number of eggs per female is 50. Mallards produced from eggs received from the Delta Research Station laid an average clutch of 8 eggs their first year. These eggs were smaller than the regular "wild" game farm mallard eggs and had a uniform buff color.

Some well-known waterfowl biologists swear that these ducks only produce one egg every other day. From my own observations, I have found that the mallard usually skips every third day during her laying period.

If the nest is on the ground, it is usually started as a shallow scraping where the first egg or two is deposited. The eggs are covered most of the time from the time the first egg is laid, using material surrounding the nest, such as grass. As egg laying continues, the nest is built up with more and more of the surrounding material. Large quantities of down added to the nest indicate that incubation is about to start or has started. The female usually sits fairly tight during the first 3 critical days of incubation. She spends more time off the nest preening, eating, and exercising about the 22nd day than the other days when she leaves the nest only for food and water. She then sits tight again for the last three days. The incubation period is 25½ days.

Pintails

The pintail nesting season in our part of the country is from April to mid-June—seldom into July. The average clutch consists of 8 or 9 olive-brown to olive-green eggs. The pintails observed followed the mallard pattern of nest building but started adding down to the nest earlier, usually about the mid-point of the laying period.

These birds followed the same pattern as most waterfowl—sitting tight for the first 3 days and off the nest for the longest and most active period about 3 days prior to hatching. The incubation period is 23 days.

Redhead

Redhead nesting in our part of the country usually starts the first of May, occasionally as early as mid-April, and continues to mid-July. If a number of females are in the same pen the average clutch size is hard to determine, as 2 or more females may deposit their eggs in the same nest. They will also deposit their eggs in nests of other species, sometimes driving an incubating female off her nest. I once observed a redhead evicting a mallard that was incubating eggs in a box set out over the water. I believe the average clutch size of a female that has decided to lay in her own nest for incubating, without additional eggs from other females, is 12. The eggs are highly waxed and cream-colored, slightly larger and more rounded than the mallard egg. Heavy downing of the nest indicates that incubation is about to begin. The redheads use much more down than the mallards or pintails. Incubation period is 26 days.

We have had 2 nestings of 2 Cuban tree ducks with 11 eggs in each. These eggs were white and fairly oval. No down was deposited in the nests either prior to incubation or during incubation. Several authorities I have consulted on these birds report that they have not seen down in the nests of any of the tree ducks.

I have observed egg turning by swans and geese, and I assume ducks do it in a similar manner. I do not believe the birds are clock-watchers and turn their eggs every 1, 2, or 4 hours. I think that some days a duck may turn her eggs 10 or 12 times and other days maybe 2 or 3 times. Among the swans and geese that I have observed turning eggs, the female would stand up in the nest and reach down over the egg, so that the top of the bill is on the ground and the mandible is hooked under the eggs. When she pulls her head up the egg is turned toward her feet and the center of the nest. She does this to each egg, turning around in the nest as she does so.

In wetting eggs, I have noted many times, especially with redheads, that the eggs were quite wet each time the female left the nest after laying her next egg. There is still disagreement as to moistening eggs during incubation. I have observed both swans and ducks complete their preening over their nests, possibly dropping moisture at the same time. I have also observed them returning to their nests, completing the preening before their return.

ROBERT V. BAUER

Bantam Chickens for Hatching Ducks

Common Eider Nesting

Silkies or silkie crosses make fine setting hens to use on duck eggs, but any bantam sets well. The commercial laying breeds have had the broodiness bred out of them and are useless.

After the hen lays a cycle of eggs, she becomes broody and will cry out in annoyance if you put your hand under her. She now may be used for hatching the eggs of other species. Eight to 12 mallard eggs or 18 wood duck eggs may be set under her, depending on her size and the size of the eggs.

Make a box 12″ x 12″ x 12″ with a removable top and a 4- to 5-inch hole in the side of the box 3 inches from the floor. Put in a 1-inch layer of dirt, rounding up the corners, then add 2 inches of bedding, shavings, short straw, grass clippings, dry leaves, or similar material. Shape this material into a nest with your fingers, then add some dummy eggs such as chicken eggs or even golf balls. Then place this box inside a cage.

The cage should be made with ½-inch hardware cloth, 3 feet long, 2 feet wide, and 2 feet high. If larger mesh is used, predators may kill the hen or steal the eggs.

After dark, place the bantam in the nest box and lock her in by covering the hole with wire or a sliding door. Do not risk your good duck eggs under her yet. After she has been setting quietly for 24 hours, remove the door to let her come out at will for feed and water. A young hen may need a little guidance in finding the entrance to the nest box, but soon she will come off the nest twice a day to stretch and feed and return quietly by herself. Some quiet handling is good at this point and she soon becomes used to being stroked. On returning to her nest she will turn the eggs with her beak and settle down upon them.

When, after 48 hours, you are sure she is setting perfectly, remove the dummy eggs and replace them with your duck eggs. Some operators turn the eggs themselves in case the bantam misses some, but usually they will do a good job. Twice a week, take her off the nest and sprinkle the eggs with warm water. If the humidity is high, less sprinkling is necessary; if the humidity is low, wet down the bedding material including the layer of dirt at the bottom. Check the eggs frequently for the presence of mites. Malathion powder is safe to use on eggs as a control if mites are present. If mites are present in any number, the hen will soon get so uncomfortable that she will break up the nest.

If one is short of setting hens, it is possible to remove the ducklings on hatching and set another clutch. Some silkie hens, properly cared for, have been known to set through several successive incubation periods.

Some bantams do not like the sounds of certain duck eggs when they hatch. Old squaw ducklings squeal and squeak and sometimes the bantams pick at the egg and inadvertently destroy them. Most bantams, however, will take loving care of the brood, teaching them to eat and drink and keeping them warm. Let them remain under her for some time until they become nest ripe or used to her, then, after she has taken them out for feed and drink, you may, if desired, remove the nest box. At this time give her game-bird starting crumbles and water in a shallow chick waterer.

Remember that baby ducks hatched under a bantam or in an incubator will drown if they get into a container of water they can't get out of. Only baby ducks hatched under a duck will absorb enough oil from the feathers of the parent to be waterproof and able to go right into the water and stay dry. Addition of substitute oils does not seem to help them.

After ducks are a few days old, give them a shallow pie pan for water. In a few days they will preen themselves, using their own oil from an oil gland located above the tail, and will soon be able to float above the water. Anyone inventing an oil suitable for waterproofing ducks could do quite well with it.

In case your bantam starts laying while brooding the ducklings, gather the eggs or she will start to set on them.

By this time your ducks should be feathered out and able to care for themselves except during severe changes of weather, when they should be kept under shelter.

KENTON C. LINT

The Artificial Incubation

Pink-Eared Duck

The hatching of eggs by artificial means is not a new practice. Certain species of birds in the family Megapodiidae experimented with artificial methods long before man tried to hatch eggs in an incubator. Even now, these birds dig holes in the sand on a beach on remote islands as they did a hundred years ago, cover the eggs with sand, and allow the sun to perform the task of incubation. Brush turkeys pile up large debris of humus and bury their eggs in it. The decomposition of the humus supplies sufficient heat to cause the eggs to hatch.

Credit for the development of the first artificial incubators must go to the Chinese and Egyptians, who have used them for centuries. Many of these primitive hatching units were indeed very crude compared with our modern incubator, and their operation was a difficult art handed down from generation to generation. They had, for instance, no thermometer for accurate control of temperature. So the early Egyptians relied on the senses of the operator to determine if the heat within the incubator was satisfactory.

The first successful incubators were developed in the United States during the latter part of the 19th century. They proved a success because man had learned how to duplicate natural conditions artificially by providing *proper temperature, humidity, ventilation, egg turning,* and *egg testing.*

Today artificial incubation is a science and an art and should be a challenge in the great field of commercial and ornamental game breeding around the world. In the United States and Canada there are many manufacturers and models of incubators to fit different needs. However, it has been found that the operator of any machine determines its efficiency. Training and experience are necessary; and I would list *patience, attention to details,* and *skill* as prime requisites for an operator.

It is wise to follow the manufacturer's recommendations with each machine because much research has gone into the building of the modern incubator. Experimentation is not advised until the manufacturer's instructions have been given a thorough trial.

As this book is written for the beginner as well as the experienced waterfowl breeder, let us review the essential parts of an incubator. There are seven essential parts:

1. The Egg Chamber
2. The Heater
3. A Temperature Regulator
4. A Humidifier
5. A Ventilating Device
6. A Turning Device
7. A Thermometer

Egg Chamber

The eggs are placed in a compartment known as the egg chamber. This should be a compact unit and well insulated so that a uniform temperature can be maintained. In the small and sectional type machines the eggs are placed on their sides on the egg trays, while in cabinet incubators the eggs are placed on their small ends in an upright position.

The Heater

Some device must be provided to furnish heat to the egg chamber. In very small incubators a lamp or electric light bulb may be used. The burners should be kept free from dirt, and after each hatch they should be thoroughly cleaned by boiling. If the hatching eggs must be turned by hand, always do this before filling the lamp. If oil comes into contact with the eggs that are hatching, a poor hatch will result as oil closes up the pores in the shell.

Large, sectional type machines are usually heated by hot water, each deck being equipped with a heating system. Oil, gas, electricity, or coal are possible sources of heat. The heaters should be cleaned frequently and the flues and pipes checked regularly so that source of heat does not escape.

The cabinet type incubators are heated by electricity. The heating elements should be tested before the incubator is operated. The switch that makes and breaks the electric current must be kept free from dust and dirt. During the hatching season most operators have a gasoline-operated generator as a standby for any emergency in case of power failure, day or night.

Temperature Regulator

Success depends upon the right temperature during incubation. A device known as a thermostat makes even temperature possible within certain practical limits. Two general types of thermostats are in use. One is a small metal disc filled with ether which expands readily when heated. The other is made of two or three metals, usually brass, steel and zinc, which expand at different rates when subjected to heat. In both cases, the force developed by the expansion pulls on levers or furnishes contact with electrical current that operates the damper or heating element which regulates the supply of heat.

Should the temperature within the egg chamber drop, the source of heat is reactivated by the action of the thermostat. If the temperature within the egg chamber becomes too high, the heat is checked by the action of the thermostat. Take plenty of time to warm up the incubator. A new machine, or one which has been idle some time, will take a few days to dry out, warm through, and adjust properly. Be sure that the regulator has been adjusted and the machine running properly for 48 hours before eggs are put in. Unless the incubator has been properly warmed up, regulator adjusted, and care taken to see that all is in right order, you risk the loss of an entire hatch.

The location of the incubator thermometer is important. Every incubator maker gives positive instructions for locating thermometers and the height of the bulb from the bottom of the egg tray. If a new thermometer is substituted for an old or broken one, be sure and get it in the right place.

A "suspended" thermometer is one that hangs over the eggs. Depending upon the type of tray and incubator, it may be located above the fan in the forced air type of incubator or near the center of egg chamber with the bottom of the bulb 1½ or 2 inches above the bottom of the egg tray.

A "contact" thermometer is one with the bulb resting on the eggs. It is usually placed near the middle of the tray with the bulb in contact with two fertile eggs, close together, about midway between the large and small ends of the eggs.

Humidity

Most incubators have some sort of arrangement for supplying moisture to the egg chamber. Water pans are placed under the hatching eggs in some machines. In others, more elaborate automatic humidifiers are installed. The use of moisture within an incubator prevents too rapid evaporation of moisture from the egg contents. Incubators do not hatch well in hot dry rooms; they do best in well-ventilated cellars or rooms free from strong drafts, where the air contains considerable moisture.

Ventilation

We have learned that a developing embryo requires fresh air. All machines have a device that can be regulated to provide the proper amount of air. As the hatch develops, more fresh air is needed. Fresh air is brought into the incubator through inlets, the number and location varying with the model of the machine. Outlets are provided for used air. Fresh air is circulated in cabinet machines and glass incubators by means of electric fans and enters the egg chambers through ventilation openings.

Air contains about 21 percent oxygen; and without adequate ventilation it is very easy to have a deficiency, because the carbon dioxide is produced at the expense of the oxygen. An excess of oxygen is not nearly so detrimental as is a deficiency. A deficiency of 5 percent oxygen reduces a hatch by about one third, or from 81 to 55 percent. If the carbon dioxide in the air of the incubator goes above 150 parts in 10,000 there will be a high mortality of the chick embryos, and this dying will be greatly increased when the carbon dioxide rises above 200 parts (or 2 percent).

Experience and recent tests have shown that when the percentage of oxygen is held constant at 21, along with constant temperature, humidity, and air movement, the decrease in percentage hatch is proportional to the increase in CO_2. A maximum CO_2 content of 0.5 percent has been worked out for optimum hatching results in many research centers.

Egg Turning

In the small type incubator, no turning device is provided, so the eggs must be turned by hand. Sectional and cabinet type incubators are fully equipped with a mechanism which turns all the eggs in the machine in a few moments. While the eggs are being turned the doors of the egg chamber should be kept closed.

Begin turning the eggs on the second day. Turning should be done carefully without shock or jar. Turning helps to equalize the distribution of warmth through the eggs and prevents the yolk and embryo from sticking to the shell membranes. The usual practice is to turn eggs in the incubator 3 or 4 times during each 24 hours, but experience has determined that increased turning, up to 8 times daily, will improve hatchability.

Egg Candling

Modern high-power transformer-type candlers and egg testers have been developed to permit detection of unproductive eggs. During the incubation period, eggs usually are candled for infertility or dead embryos on the 7th and 14th days. Eggs in which the embryo has died may be replaced on the 14th day of incubation.

Waterfowl Incubation

The incubators are operated at a temperature of 99½° F. and a wet bulb reading of 55 percent at all times. The eggs are removed from the incubator and placed in the hatcher on the morning of the 24th day. The hatcher is operated at a temperature of 99° F. and a wet bulb reading of 65 percent on the 24th day; 70 percent on the 25th day; 75 percent the 26th day and 85 percent–86 percent the 27th day. The ducklings are removed from the hatcher on the morning of the 28th day. With other species of waterfowl with longer incubation periods, the same pre-4 day schedule may be adjusted in artificial incubation. Fifty-two species of waterfowl in the collection have been hatched successfully in incubators.

JOHN ELLIS

Using an Artificial Incubator

Using incubators to hatch waterfowl eggs can be very successful, and I recommend it to anyone who rears large numbers of birds. Unfortunately there are no set rules for success with incubators, as climate, elevation, air pressure, and probably many other factors influence the success or failure of artificial incubation. I have been most successful with forced-air incubators and have used them exclusively in hatching all of my waterfowl eggs during the past 10 years. I have successfully incubated and hatched swan, goose, and duck eggs in the same incubator at the same time with 90 to 100 percent hatch of fertile eggs. In theory this is impossible—that is why I feel that location is a big factor in success or failure.

For the person who is interested in trying artificial incubation I would suggest he write for catalogues from each of the better known manufacturers. After careful study of these catalogues he should pick out an incubator that he feels best serves his purpose. My only recommendation is that one must have at least two incubators, one to be used for incubation, the other for hatching.

There are two types of incubators: The still-air type and the forced-air. I would not recommend the still-air as they are much more difficult to control and you cannot hatch eggs of different sizes at the same time.

Each incubator manufacturer has his own suggestions and rules which in substance vary somewhat, but they admit that much common sense and trial and error are the only rules for successful hatches.

From my own experience I offer the following information which might be helpful.

Humidity is undoubtedly the most difficult factor to control and very important for successful hatches. As a guideline the relative humidity for duck eggs should be between 85–90 degrees during incubating and 90–95 during hatching time. Goose eggs require 85–90 degrees for the first 25–26 days and 90–95 for the last 4 days. Some of the smaller species of geese which hatch after 22–24 days' incubation must have higher humidity than the 18–20 days.

The best way to determine whether the moisture inside your incubator is right is to candle the eggs frequently. If the eggs are drying down properly the air space should be ¼ of the size of the egg. At hatching time it should be ⅓ of the egg space, a little more with geese and swan.

There are 3 ways to control the humidity in your incubator. The first is your moisture pan, which is usually sectioned into two parts. Some locations will require both to be filled while others the larger or smaller section. The second is by opening the air vents in the machine. The larger one makes the vents the more humidity is allowed to escape. The air vents are very important and probably the hardest to adjust properly as outside humidity is an important factor and can vary greatly from day to day or week to week. I would recommend that any person who is going to incubate eggs for the first time purchase his incubator a month or two before he plans to use it. Run your incubator and experiment with the water pans, vents, etc., using a wet bulb hygrometer as a guide. When the eggs are finally placed in the incubator you will have some knowledge as to how it must be run.

Temperature is the other factor which must be controlled, and the thermostat may have to be changed occasionally because of changes in room temperature and humidity. Incubators must be placed in a cellar or room at 60 to 70 degrees. The room used must not be one with great changes of temperature as this affects the inside temperature of the incubator. Place your incubator against a wall that is free of draft. Do not place it where direct sunlight shines on the incubator. If the room is very dry use a humidifier or open pans of water to increase the room humidity. Ideal room humidity is 60 to 70 degrees.

The correct operating temperature for each type of incubator has been figured out by the manufacturer. Follow his instructions and you won't go wrong. Different types of incubators require a different temperature depending on where the thermometer is placed.

Waterfowl eggs must be laid flat in the egg traps and always turned to a full 180 degrees. The eggs must be turned a minimum of 3 times a day. Turning eggs 5 or 7 times a day is even better if it is possible, but good hatches can be expected if only 3 turns a day are made, providing heat and humidity have been correct. It is advisable to cool the eggs every second or third day. Unless the humidity has been too high a sprinkling with cool water is advisable just before the eggs are returned to the incubator. The cooling time should be 10 to 20 minutes at 60- to 70-degree room temperature, less if cooler.

When the eggs have been pipped it is time to place them in the incubator that is to be used as the hatcher. From this time on the eggs should not be turned. The incubator should be opened only twice a day during this period. The eggs should be sprinkled or sprayed with warm water twice a day until the ducklings or goslings begin to hatch.

There can be a great variation between pipping and hatching so one should be patient and wait and not try to help the birds hatch. For the most part if a bird isn't strong enough to hatch on his own he will usually die later on. I've seen birds hatch only a few hours after pipping and have had birds hatch that were perfectly healthy and sound 4 and 5 days after they pipped the egg. Some birds will hatch a day to three days later than others in the same clutch of eggs. For the most part you will kill some young ones trying to help them out of the shell that would probably have made it on their own. I am convinced from my experience that if a duckling or gosling isn't healthy enough or strong enough to hatch on his own power he would probably never grow to maturity.

GUY A. GREENWELL

On Helping Ducklings Out of the Egg

Hatching wild duck eggs, for the average breeder, is not a mass-production operation. If he operates an incubator, he sets eggs of several kinds and usually has eggs in all stages of incubation in the incubator together. As a result, he has no opportunity to learn the ideal levels of temperature and humidity for the eggs of each species. He could not achieve them, if he knew them, in the average operation.

If his experience parallels ours, some ducklings will pip but go no further. Some will have died at some point during their emergence.

Can those ducklings be helped and still emerge as healthy individuals? Our experience indicates that they can. My wife hit upon a significant ingredient in doing it some years ago. We keep a laundry sprinkler-bottle of water in the hatcher and thereby have a supply of water at the correct temperature at hand. Another essential element is patience, as will be indicated by a description of our procedure.

The egg shell is lined with a heavy membrane and inside that the bird is encased in a much more delicate membrane filled with blood vessels. It is that inner membrane which forms the lower surface of the air cell if the egg is held large end up. We hold the egg in that position near a strong light, which serves to warm the chick, if we have to work on it. And we close and restart the incubator while we work on each egg. Thus, no others are chilled.

In pipping, the bird's beak punctures the inner membrane and extends into the air cell. Very soon thereafter, it must puncture the outer membrane and begin to breathe outside air. This seems to give it the stimulus and the strength to turn in the egg and cut its way around. And, if it is able to make that struggle, the absorption of the yolk and the passage of the blood supply in the membrane into the body proceeds quite rapidly.

Because of other demands upon my time, we turn and inspect the eggs 3 times daily, at 8-hour intervals. Our discussion is based on that regimen.

If an egg has pipped and made no further progress within 16 hours, we take the first steps. We puncture the membrane lining the shell at the pipping point with a sharp pencil. We hold the pencil at an angle, pointed toward the large end, and away from the bird's bill. We lift the pencil point, making a small tear in the membrane. We make sure the membrane does not return to close the opening, even removing a tiny piece of it to admit a bit of fresh air.

Then, beginning a bit to the side of the opening we have made, and following the path we imagine the bird's bill would traverse if it turned normally and cut the shell, we make a very narrow crack in the shell about two-thirds of the way around. We do this with a thumb nail. We do not begin at the pipping point for we would be apt to create a diffuse crack or "cave-in." We do not attempt to cut the outer membrane. If there is a "cave-in," the bird will not be able to cut past it. If the outer membrane is penetrated at that level on the egg and at that stage of development, fatal bleeding will result.

We then return the egg to the hatcher. When we inspect the eggs after another 8 hours, a fair portion of those we have treated will have emerged without further help.

Those which have not progressed are in real trouble, and additional aid must be administered with great care. We enlarge the opening at the pipping point, again working toward the large end and avoiding the inner membrane. We make a near circular opening large enough to let us view the inner membrane. We sprinkle that membrane with a few drops of our tempered water, taking care to avoid flooding the bird's nostrils.

The membrane will have dried, becoming whitish in color, but when it is sprinkled, it becomes pinkish and we can see the blood-filled vessels. At this stage, the blood will have regressed from only a small area of the membrane, right around the bird's beak. With the pencil point, we carefully lift the flap of the membrane atop the bird's bill and try to push it back so it cannot dry across the nostrils.

We know that without the violent exercise of the normal turning, the blood in the membrane and the yolk will be taken into the body very slowly, so we return the egg to the hatcher for another 8-hour period.

We examine the membrane each 8 hours, sprinkle the membrane liberally, and again make sure the nostrils are clear. Each time, the membrane will have dried, and the bird will have become quiescent.

When it is moistened, however, the bird will become active and even drink eagerly if we get a drop of water on the bill.

After 8, 16, or even 24 hours, we find that the membrane has turned yellowish in color and appears shrunken. It retains its yellow color when we moisten it. This is our indicator that the blood supply has receded from it and we may safely tear it. We enlarge the opening in the shell and examine the membrane to the very edges of the air cell.

The blood supply recedes from the membrane and it becomes yellowish beginning on a line from the tip of the bird's bill to the center of the large end of the egg. Often, we can begin removal of the membrane in that area and turn it back over the edges of the shell to prevent its sticking to the bird. Then we must wait until the blood has receded further before we can safely disturb the remainder. If it sticks to the bird, a bit of the warm water quickly softens it so we can lift it off with the point of our pencil. We remove more as we can.

When the membrane has lost its blood supply down to the crack line we created around the egg earlier, the bird will kick out the shell after we've removed the shell, turned back the membrane, and returned the egg to the hatcher. We never pull a bird from the shell, for some will not have absorbed the yolk completely, even at this stage. We give them a bit more time, still sprinkling them liberally.

A very few individuals are too weak to kick out but can still survive. We help them by lifting the exposed wing, again with the pencil point, so it is free. Very rarely, we have to help a bird more by lifting its head free. We do this by beginning at the base of the neck and gently lifting and unrolling the neck with thumb and forefinger.

Those slowest birds usually gain strength slowly and seem to dry more slowly than normal. However, if we have been able to go through the entire process of helping them without causing any bleeding, most of them are indistinguishable from normally hatched birds within a couple of days.

Even the loss of a small amount of blood is not fatal if the damaged area is immediately sprinkled generously with the warm water. We've had a couple of instances of apparent brain damage, but those are obvious. We have been unable to find any evidence that the others do not become normal adults. It has been about like trying to pick out the "incubator babies" among a group of human adults.

Dr. William Ferguson adds these comments: "As I read my notes at the end of my hatching season this year, I made the following observation: 'When a duckling pips and fails to make progress for a number of hours such that the shell around the initial two or three pips begins to have a discolored or dried look, it is time to make a small air hole through the outer shell membrane. The presence of the air hole seems to stimulate the further activity of the duckling and to preserve his life. Most birds handled in this way seem to come through in good shape without further assistance.'

"I wonder how many people know this little trick of making the air hole? Next season I will certainly try again, after Greenwell's more complete description, the method of using a wetting solution. Because of my own work with eye surgery and the damage to tissues that ordinary water (even sterile), and to a much less extent normal saline solution does, I use Ringer's, which is a balanced physiological solution and does not harm the tissues. You might consider this as a replacement solution for your water in future years."

G. MICHAEL FLIEG

Incubators and Methods

1. HUMIDAIRE MODEL 14 GOOSER for waterfowl eggs. Capacity—750 duck eggs or 300 goose eggs; 180-degree automatic turning every hour.
2. HUMIDAIRE MODEL 550 HATCHETTE for game birds and ratites. Has separate incubating and hatching units. Capacity of the incubating unit—400 chick eggs or 15 ostrich eggs or 400 pheasant eggs. Hatcher has capacity of 150 pheasant eggs; 45-degree automatic turning every hour.
3. HUMIDAIRE MODEL 10. Very small incubator. No automatic turning but useful for eggs in which humidity falls between the other two incubators. Capacity—100 pheasant eggs.

Air-conditioned area is best for incubators because the cooler the outside air the more oxygen is brought into the incubator.

Clean incubators as much as possible during the hatching period with a good chlorinated disinfectant and fumigate with potassium permanganate, ammonia, and 10 percent formaldehyde solution. This can be done, if no eggs are pipped, with no mortality. Clean eggs with a commercial egg-cleaning solution before setting in incubator to minimize disease. Exception—game birds' eggs.

NOTE 1. Measure your doors and be sure to get complete measurements of incubators *before* ordering them, as they may not fit into your building!

NOTE 2. Keep a supply of switches and indicator lights on hand, as these may go out during the middle of the hatching season and you could lose the whole hatch this way!

Incubation temperature: 99½ to 99¾ degrees F. for all birds but ratites; 98 to 99 degrees F. for ratites.

Incubation humidity: 88 to 92 percent waterfowl and gulls.
86 to 88 percent cranes.
83 to 85 percent pheasants, quail, grouse, other game birds.

Hatching by incubator: All eggs come out of incubating trays and into hatching trays 4 days prior to hatching.

Rearing birds: I use Jamesway Battery Brooder with heatpad-type unit where birds stand beneath and warm their backs.

Start birds with unmedicated Purina Startena or Purina Game Bird Crumbles with LIV sprinkled on top. Divers receive one-quarter Trout Chow #4. Greens to be fed in A.M. I recommend starting waterfowl on a wet mash, pheasants and game birds on a dry mash—but always a dry mash overnight, as wet will spoil. Geese go on Chick Growena when taken from brooder into rearing pens. If wings start drooping, clear this condition up.

Rear birds on wire, if possible, but take waterfowl off wire after they lose their down or foot problems may develop. Put them in a cage with a shallow pool and sandy loam type soil for sanitation.

WILLIAM HUEY

Specialized Incubation

Even though the subject of incubation has already been covered by other writers, I would like to suggest a specialized incubator for those aviculturists of the Rocky Mountain region or other areas where altitude is a problem in the hatching of eggs. It has been demonstrated that the oxygen requirement during the hatching period is critical in high altitude areas.

I tried a number of commercially built incubators with varying degrees of success before I finally decided to construct a machine designed as nearly as possible for the conditions which existed at our home, located at an elevation of about 7,000 feet. It appeared that the well-insulated cabinet required for this construction could be more easily obtained by the conversion of a refrigerator than by starting from scratch and building one.

I searched around the appliance dealers and found a small, apartment-sized refrigerator which I bought at minimal cost. The bottom portion, which had housed the refrigeration mechanism, was removed with a hack-saw and wooden blocks were attached to serve as legs. Since the problem to be solved involved the oxygen supply, the design centered around the ventilation ports. Consideration was given to the location of these ports so that they might contribute to an even distribution of air throughout the entire machine. The intake port was located near the bottom of the machine on the right side immediately below the place selected for the heating element. The discharge ports were located near the upper left-hand corner of the machine, opposite the fan. Before the complete design of the machine was finished, the efficiency of the location of these openings was tested with the fan and divider partition in place by injecting smoke into the intake and observing its circulation and discharge. In order to assure a rapid turnover of air and a generous supply of oxygen, a 2-inch diameter was arbitrarily selected as a size for the ports. Adjustable covers were provided so that the humidity of the air inside the machine could also be controlled.

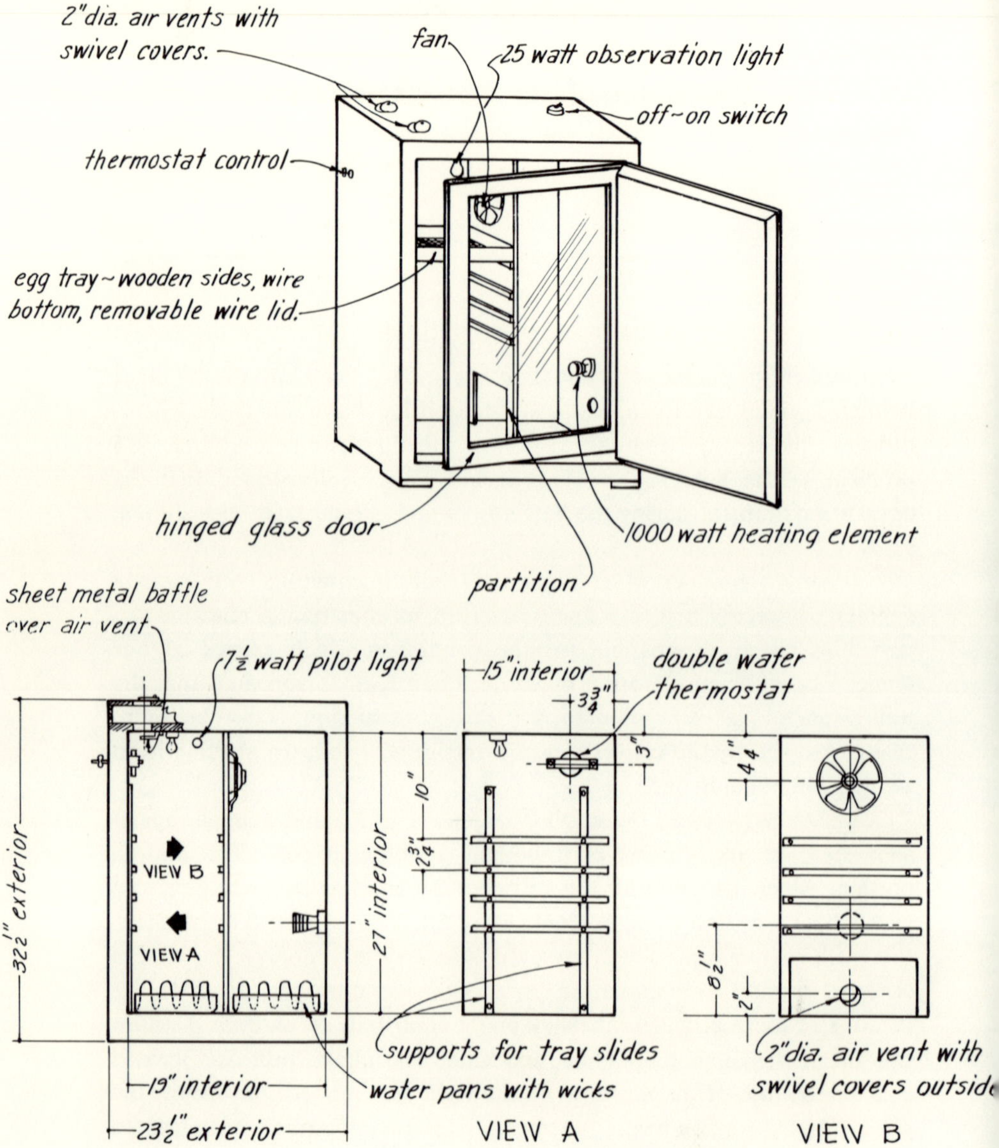

SPECIAL HIGH-ALTITUDE INCUBATOR

Special High-Altitude Incubator

MARY B. HUEY

A diagram of the design of the machine is included which shows the location of the divider between the egg tray compartment and the heating compartment, the location of the fan heating element, thermostat, pilot light, observation light, etc. In order to accommodate the rapid exchange of air in the machine, and to guarantee as nearly as possible a uniform temperature, an over-sized heating element is used. From a trial and error process, a 1,000-watt coil-type element has been selected as the most desirable. A glass door was installed inside the insulated door so that observation during hatching could be made without disturbing the atmosphere inside the incubator. This has proved to be a very desirable item. This machine is also wired in such a way that the observation light inside the machine is actuated by the automatic switch which turns the light on in most ordinary refrigerators when the door is opened.

The egg trays are designed in such a way that a cover is required on only the top tray. The space between the lower trays is small enough to prohibit hatched chicks or ducklings jumping from the tray. Humidity in the machine is maintained by manipulation of the air ports and is supplied from water in two galvanized metal trays that completely cover the floor of the machine. In order to increase the evaporation from these trays, a system of wicks was included that contain cotton flannel cloth sewed to a series of wire supports in an accordion-like arrangement. The eggs are placed loose in the tray on their sides and are turned periodically by hand.

The electrical components used in the construction of this machine are available from Lyon Rural Electric Company, 3425 Hancock Street, P.O. Box 30, San Diego, California 92112. The double wafer thermostat is recommended as a safeguard against one faulty wafer. The positive action separation to break the electrical connection is also recommended over the snap-switch type used in some thermostats for the same reason, as a safeguard against a faulty switch.

A Hatch of Wild Mallards

An Attractive Layout

J. J. SCHEDEL

Waterfowl Pens

J. J. SCHEDEL

An Aviary Interior

The Beauty of Ponds and Landscaping

PHOTOS: J. J. SCHEDEL

An Elaborate Bantam House

A Nest Box on a Lily Pond J. J. SCHEDEL

CARL STRUTZ

Pintail Drake. Note Screening to Protect Banks

ED BRY

Gadwalls in Winter

Feeding Time

Living with Waterfowl

J. J. SCHEDEL

Open Water in Winter

GEORGE MILLER

Keeping Water Open with Bubblers

Shoveler

ED BRY

ED BRY

Pintails and Widgeon

Privacy through Natural Vegetation

CALVIN WILSON

OAKLAND PARK DEPARTMENT

Feeding Time at Lake Merrit Waterfowl Refuge, Oakland, California

A Nesting Mallard and Friend

SAN DIEGO ZOO

DAYTON O. HYDE

Canada Goose Tending Eggs

A Hidden Nest

THE FLEISHMAN CARIBBEAN GARDENS

Redhead Nest Hatching Out

ED BRY

A Hatching of Pochards

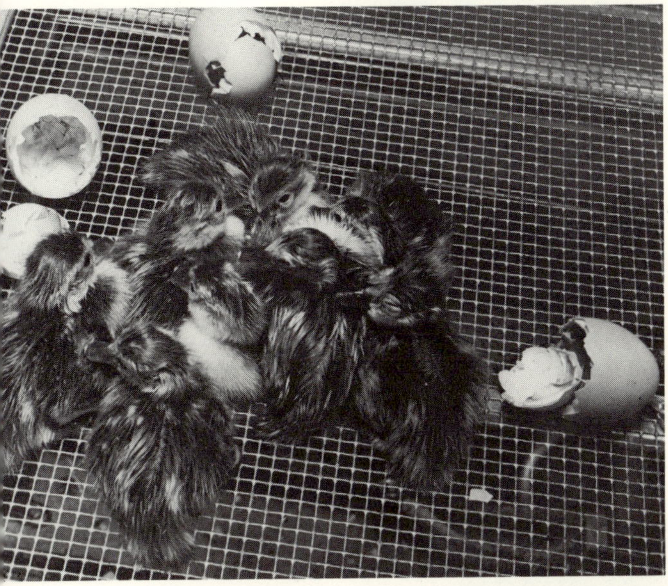

THE FLEISHMAN CARIBBEAN GARDENS

Young Mallards

ED BRY

Shovelers ED BRY

SAN DIEGO ZOO

Mandarin Ducks

Ringed Teal

RON GARRISON, SAN DIEGO ZOO

Ringed Teal Family

PAUL JOHNSGARD

PHOTOS: ED BRY

Mallard and Leg Band

Holding for Banding

Caught for Banding

Proper Placement of Band

Newly Banded Baldpate

Releasing After Banding

Steller's Eide

PHOTOS: PAUL JOHNSGARD

Buffle-Heads

Sneak Display of Male Canvasbac

Ruddy Duck

Bronzed Winged Duck. Male Displaying

PHOTOS: PAUL JOHNSGARD

PART V

Brooding

FORREST B. LEE

Artificial Brooding of Ducklings

Wood Duck at Nesting Hole

The brooder is a facility designed for starting out in life newly hatched ducklings which must learn to eat, drink, and live in a captive environment. It is a home or nursery for ducklings reared artificially without a mother or foster parent. It supplies food and water and the warmth which is ordinarily provided by the hen during the first days of life.

There are many kinds of brooders; however, the infrared heat-lamp electric type is probably the most satisfactory for starting ducklings of wild duck species. It is versatile and can be adapted to many different situations and conditions. It is especially useful where small, separate groups of ducklings are being handled. For example, a simple infrared brooder can be made by suspending a 250-watt heat lamp over a relatively small floor space. A circular guard of corrugated cardboard or tin 12 to 15 inches high is placed around the area of the heat lamp at a distance of about 2½ to 3 feet from the center of the heated zone. Litter is spread on the floor within the guard and the heat lamp is adjusted so its face is about 18 inches above the litter. Water is provided in ordinary chick waterers, and starter mash is put in saucers or flat pans.

While this simple arrangement is satisfactory for starting some species, many propagators use more refined infrared brooding methods.

These usually consist of a brooder box with a hardware cloth bottom and equipped with heat lamp, waterer, and feed pan. The screen bottom permits better sanitation and is generally more satisfactory than using litter. Some propagators prefer giving newly hatched ducklings access to water to swim in and use a shallow pool arrangement in addition to the infrared lamp and feeder. The pool should be placed away from the heat lamp so that water droplets are not splashed onto the bulb causing it to break.

Figures 1, 2, 3, and 4 illustrate a type of infrared brooder used with good results at the Northern Prairie Wildlife Research Center. Species successfully started in it include mallard, pintail, black duck, gadwall, baldpate, blue-winged teal, green-winged teal, redhead, canvasback, lesser scaup, common goldeneye, wood duck, mandarin, and ruddy duck. Up to 15 ducklings can be handled satisfactorily in this brooder at one time.

The brooder box (Figures 1 and 2) is 22″ high, 40″ long, and 26″ wide. It is made of 1″ boards or plywood at least ⅜″ thick. It is supported by four 2″ x 2″ legs or posts 36″ long which are attached on the ends of the box. The lower edge of the box is 14″ from the floor without casters. Casters are recommended, however, since they facilitate moving and cleaning the brooders. The bottom screen (Figure 3) is of ¼″ hardware cloth attached to a frame 24″ x 38″. This unit fits into the brooder box where it is supported by a wood ledge (Figure 2) of 1″ x 1″ material 3″ above the lower edge of the box. Additional support under the screen for the waterer and feeder is provided by two cross pieces of 1″ x 1″ wood 6″ apart and near the center of the brooder (Figure 4). A 2″ x 2″ post (Figures 1 and 2) is attached in one corner of the brooder box. This post projects about 14″ above the level of the box and the upper end has 3 or 4 ¼″ holes drilled diagonally across toward the center of the brooder (Figure 4). The end of an 18″ length of ¼″ steel rod with a hook on one end is inserted into one of these holes to provide a support for the heat lamp. The height of the lamp can be adjusted by putting the steel rod into a higher or lower hole.

A distance of 18″ from the face of the heat lamp to the bottom screen is generally satisfactory if a 250-watt bulb is used. The temperature in the heated area below the lamp at the screen level should be in the range of 90 to 95 degrees F. for starting newly hatched ducklings. The behavior of the ducklings will generally indicate whether the lamp should be adjusted up or down. If the temperature is too high, they will

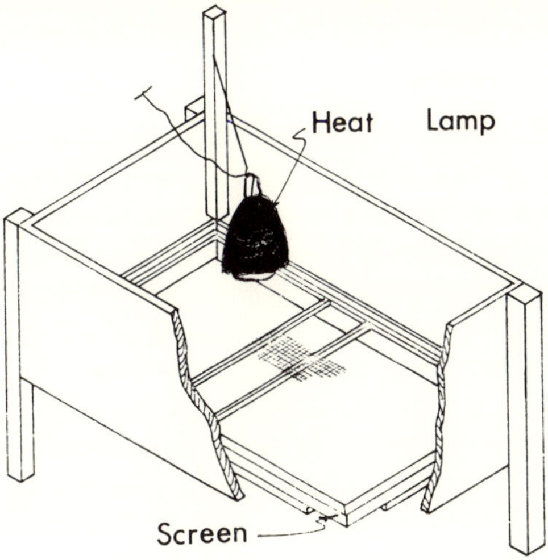

FIGURE I

View of Brooder Showing Location of Heat Lamp and Bottom Screen

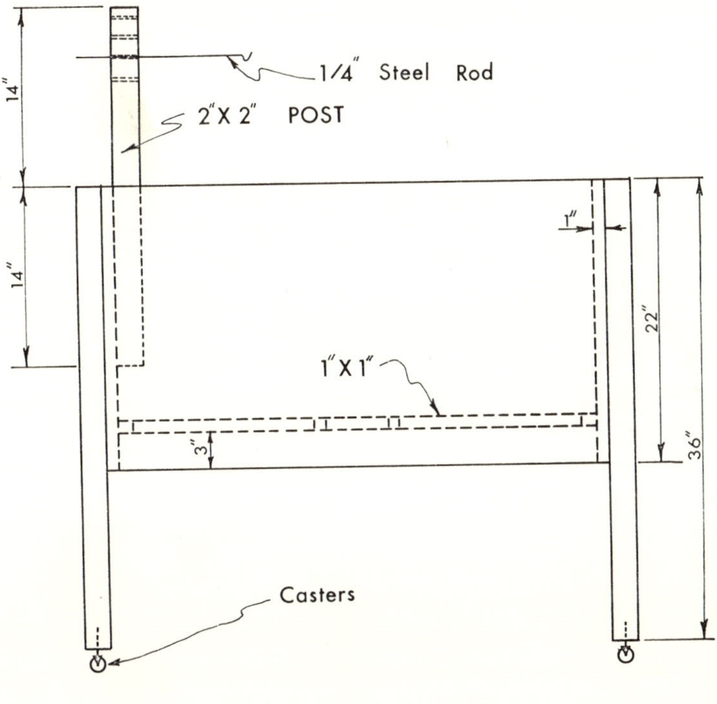

FIGURE 2

Side View of Brooder

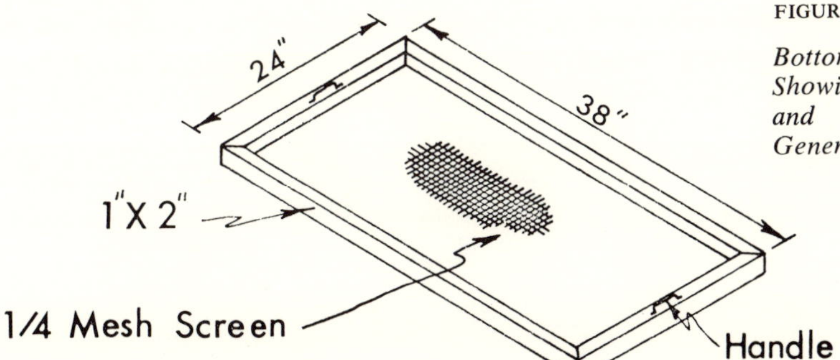

FIGURE 3

Bottom Screen Showing Dimensions and General Construction

DRAWINGS: FORREST B. LEE

crowd away from the heat. If the temperature is too low, they tend to pile up or huddle together. When the brooding temperature is right the ducklings will be well distributed over the floor or will form a doughnut-shaped circle under the heat lamp when sleeping. Infrared heat lamps come in three sizes: 125, 150, and 250 watts. While the 250-watt is most popular, circumstances may warrant the use of the others, and the distance of the lamp to the ducklings should be adjusted accordingly.

The waterer should be placed near the middle of the box so it is supported by the two cross-pieces. It should be far enough from the heated area so that the water does not become warm and cannot be

FIGURE 4

View of Brooder from Top

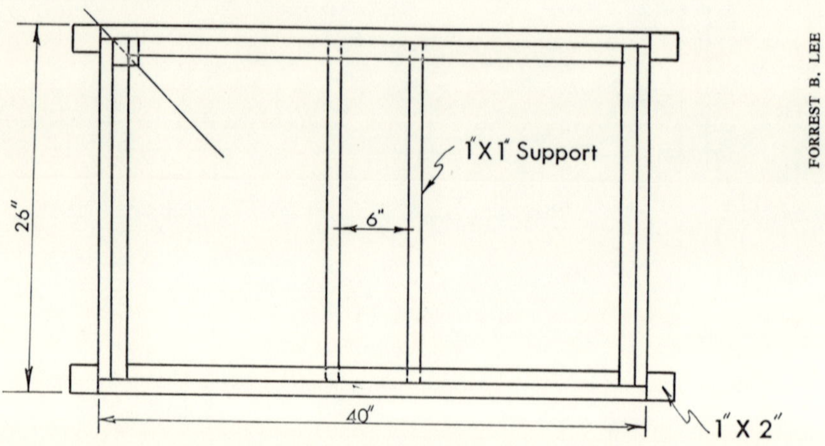

splashed on the heat lamp by the ducklings. Likewise, the feed pan should be placed back from the heated zone.

If several brooders are to be used, it is important that the design be standardized and the bottom screen units be interchangeable. In this way a supply of extra screens can be made, and it becomes a simple matter to replace a dirty screen with a clean one. Also, it is advisable to have some bottom screen units made up with ½" hardware cloth. They stay cleaner than the ¼" units and after the ducklings are a couple of days old one of these can be substituted for the ¼" bottom screen.

A brooder cover made of a wood frame and ¼" hardware cloth is used for starting wood ducks, mandarins, goldeneyes, and other ducklings which may otherwise jump out. Usually the cover is necessary for only the first 2 days until the ducklings lose the urge to jump.

Figure 5 shows a brooder with the end removed so that the interior can be viewed. This particular unit can be dismantled into sections for winter storage. The end piece being held has the piece of 1″ x 1″ wood attached which helps support the bottom screen unit.

FIGURE 5

View of Brooder with End Removed Showing Waterer and Feed Pan. End Piece Is Held with the Inside Facing the Camera and Shows Cleat Which Helps Support Bottom Screen

NORTHERN PRAIRIE WILDLIFE RESEARCH CENTER

The brooder should be readied in advance for a group of ducklings. This is possible since the ducklings are usually kept in the incubator several hours after hatching before being removed. The brooder itself and its utensils and accessories should be immaculately clean before the ducklings are put into it.

In the Northern Prairie Wildlife Research Center the bottom screen is often covered with a piece of cloth or rough paper for the first 2 days. The waterer and feed pan are put in as usual, but, in addition, some of the starter mash is sprinkled over the bottom on the cloth or paper. It is believed that ducklings of some species, such as the ruddy duck, learn to take the starter mash much more readily when this is done. All ducklings have been started on dry mash at Northern Prairie; however, some propagators wet the food.

There is much art to starting ducklings, and many propagators have developed special practices which are helpful in getting them to take food. For example, some sprinkle particles of starter mash on the backs of the ducklings, while others put in a started duckling as a teacher. Generally, however, the ducklings will start taking the starter mash quite readily.

Once started, the ducklings grow fast and soon must be moved to new quarters since the brooder becomes crowded. This move is generally made when they are from 5 to 10 days of age, depending on the species and the progress they are making. At the Northern Prairie Wildlife Research Center some are moved from the brooders to an outdoor A-pen (Figure 6) equipped with a heat lamp and a poultry waterer and feeder. The pen is constructed of removable panels, and the unit generally used has a basal area of 12′ x 24′. Once the ducklings have adjusted to the new quarters, a flat trough is added, giving them water to swim in. Other ducklings are put in pens which have concrete pools of water. Here, again, access to a heat lamp is provided at the time of the move and for a few days thereafter. The ducklings require less heat after the first week of age.

Circumstances of climate and weather, available facilities, and personal preference do much to determine how the ducklings are handled after they are started. Some propagators prefer to get them into water as soon as possible; others will rear them to adult size on wire. Undoubtedly, any of many methods can be successful if the basic rules of sanitation, nutrition, and protection are followed. Following are some

FIGURE 6

A Pen for Holding Started Ducklings. A Shallow Water Trough Should Be Provided Before Ducks Are Moved to a Pond in Order That They Can Clean and Oil Their Plumage

general considerations and precautions which may prove helpful to the beginning aviculturist.

The brooder house may be an old shed or spare room in a barn, or it might be a facility designed especially for brooders. Ideally it should have a concrete floor with floor drains and running water so it can be easily hosed down and disinfected. It should be constructed so that entry by rodents and insects is prohibited or at least limited. The ventilation system should allow a good movement of air, without drafts. A desirable brooder room temperature is 70 degrees F.; however, this is not as critical with the infrared lamp brooder as with some other types of brooders. When several infrared brooders are used in a room there is a danger of the room temperature becoming too high if the ventilating system is not adequate.

Regardless of the type of brooder used, the brooder room should have a minimum of disturbance. Ordinarily it should be frequented only by personnel feeding, cleaning, or performing other service functions.

Power failure and faulty electric circuits can put electrical brooders out of order. The circuits should be checked for faulty wiring. Be sure to use lamps and sockets intended for use with infrared heat bulbs. When power failure occurs, there may be losses caused by smothering of ducklings which pile up in a corner of the brooder. An alarm system to warn of power failure is desirable. Also, it is well to plan in advance what will be done in case of power failure or fire. A special effort should be made to reduce fire hazards in the brooder room. Care should be taken so that litter is not placed where it might be ignited by a heat lamp.

One cannot say too much about cleanliness and sanitation in the brooder room. The brooder room itself, the brooders, and the utensils should be kept immaculately clean. The sanitation program should involve a daily routine of cleaning and disinfecting. A number of good disinfectants are available for poultry houses and utensils, and one of these should be used. Regardless of the type of brooder used it is desirable to place feeders and waterers on special wire platforms since this helps keep the feed and water clean.

It is vitally important that the ducklings be started on a proper diet. If litter is used in the brooding procedure, one must be sure to avoid wet or damp litter. Shavings, ground corn cobs, peat moss, chopped straw, and commercial litter are sometimes used; however, there is danger that some species, such as ruddy ducklings, may ingest the litter, causing their death. Fecal matter should be cleaned up each day. If duckling losses occur, steps should be taken immediately to have the birds examined to determine cause of death so that corrective or preventative actions can be taken.

ROBERT V. BAUER

A Successful Method of Brooding Wild Waterfowl

In starting our waterfowl propagation under the Fried Heuser Foundation, we considered the several known methods of brooding, mainly the "dry" method used by most bird fanciers and Pine Forest Farms itself for producing mallards on a commercial scale and the "wet" method used by the New York State Conservation Department and Delta Waterfowl Research Station in Manitoba, Canada. We liked the results obtained in the "wet" method, so we scaled down the brooding facilities used in New York to a more practical scale for the fancier to obtain the same results.

We erected a multi-purpose building along the east side of one of our covered pens. This building, 130′ long by 15′ wide, houses the more delicate species during the winter and is used in part as a brooder house during the brooding season. A concrete trough was laid inside—along the pen side of the building. A conrete dam with overflow was formed at each pen partition to give us a flowing pool of water 15′ long, 12″ wide, and 8″ deep. The brooding area of sand slopes up gradually from the rounded lip of the concrete pool. Heat for the brooding area is supplied by electric infrared heat lamps. The same pump that supplies water to our breeder pens supplies the water to the inside trough through a separate pipe and valve.

The newly hatched waterfowl are confined to the brooding area for the first day after hatching. This teaches them where the heat and food are.

Some of the more difficult species to raise are started on a wet mash. Duck starter pellets are placed in the base of their drinking fountains, making a very wet mash the first day. After the first day, the guard confining them to the brooding area is opened, allowing them into their pool. The wet mash for the more difficult species is stiffened a little each day until they are weaned over to a dry ration in about a week's time.

Duckweed is floated on their pools twice daily for their supplemental feeding. Almost all species that we have raised love this feed.

We have run cygnets, goslings, and ducklings at the same time in the same brooder. They grow healthy and vigorous with this type of brooding. They are very active in swimming, diving, and running to and from the brooding area—the theory being that their oil glands start functioning right away if they swim so they stay fluffy and natural looking. The down and new feathers do not take on a dry-looking appearance as they do when brooded with only drinking water available to them.

As long as we keep the brooding area dry and the young are eating, there is little or no mortality experienced. Keeping the brooding area dry is of utmost importance. A wet brooding area will chill the birds and soak up their down. We remove the sand as it gets damp and replace it with dry sand. If litter is used, the damp or wet litter is replaced with dry litter. If a concrete brooding area is used without litter, hosing it down daily to remove the dirt will work.

When the birds are a week old, the brooding lamp can be turned off during the day, when they are two weeks old the brooding lamps are not used at all, and when they are three weeks old they can be put outside on a larger pool. When the young are moved outside to a larger pen with a larger water area, they do not soak up as the birds do when first put on the water from the "dry" method of brooding.

We have now converted the Pine Forest Farm commercial mallard brooders over to the "wet" method of brooding with very good success.

With our sandy brooding area and a shallow foundation wall, we do have rat problems. We have lost no young to the rats so far, but it is time-consuming keeping the rats down with poison baits and traps.

So far this year we have run over 200 young waterfowl from 25 species through our brooders.

GUY A. GREENWELL

Imprinting

Fulvous Tree Duck

Keepers of wild waterfowl encounter a bird now and then which will not mate with its own kind. That condition is more common among geese and swans, but it does occur among ducks. When it happens, it is usually the fault of the breeder, so he should know more about it.

The problem occurs because the bird does not know what kind it is. No bird is hatched with that knowledge "built-in." In fact, *we* are not born so equipped. To understand how the bird can grow up without learning its species, we must discard the notion that the young bird learns as we do, by a gradually developing ability to reason. The duckling never develops the power of reason, nor does any other bird, with very limited exceptions.

The duck responds to any stimulus with the reaction which has proved most likely to result in survival. Only those birds which have responded in that way have survived to reproduce. Students of bird behavior describe those behavior patterns which are performed the same way by each succeeding generation as "fixed." Each bird can perform them in only one way.

Communication among them is accomplished principally by fixed patterns. Some of them are designated as "intention movements." For instance, many kinds of ducks signal their intent to fly up with upward jerks of the head or bill. Those movements vary among species, but are easily understood by other kinds.

In order that the many species could develop and then remain pure, there had to be protection against response to the signals of other species in courting and reproduction. The duckling is hatched with the ability to be stimulated by and respond to the particular courtship displays, color patterns, and voice of its own kind.

Then, as an additional safeguard, it "learns" its own kind during and shortly after hatching, as part of the process known as "imprinting." The hatchling receives impressions of the size, shape, actions, and voices of its mother and nest mates. Within a very short time, determined experimentally as by the 12th hour or so after hatching, the duckling has been "imprinted" with those impressions and they will remain with it for life.

We can see that a bird hatched from a lone egg of its kind, in an incubator with the eggs of other kinds, could receive some false first impressions. If we expose it to a view of ourselves as the first larger object it sees, and speak to it in low and soft tones, we may find it has become imprinted upon us as its mother.

Dr. Konrad Lorenz demonstrated that he could imprint whole broods of incubator hatched mallards upon himself if he crawled about and imitated the voice of their mother. If he stood up, he was too tall to make the impression firm. However, he was not too tall to imprint greylag goslings. And I have learned that I have to be quite careful to avoid imprinting ducklings while working with them in a table-top level brooder.

When our one-of-a-kind bird spends the first few days of its life with a brood of another kind, we may find that it has become one of them. And we've had birds, reared as single young by a bantam, which never rid themselves of the impression that they were bantams.

Some birds overcome the handicap of a misguided beginning and become good and true breeders. In those cases the inborn protection mentioned earlier has proved effective. The bird can respond sexually only to the displays and patterns of its own species, though it can react to other communications from its adopted fellows.

That protection is most highly developed among those closely related species of ducks which occupy the same breeding and wintering ranges. Highly distinctive and complicated color patterns in the drakes and completely ritualized courtship behavior patterns have evolved among them. On the other hand, some geese and swans begin choosing mates while flocks of subadults and unmated birds are loafing in relative isolation from closely related kinds. Such species have not developed rigidly different courtship behavior.

In captivity, those kinds with their less effective species isolating mechanisms are more apt to go astray. The breeder must be careful in his handling of them from the egg until they are well mated. In fact, he should strive to supply all the needs of his adults. If his pairs produce full clutches, much of this trouble can be avoided.

This brief discussion has barely touched on the findings of the behaviorists. It has oversimplified most and omitted some of the interactions which make up behavior. For any breeder who wishes to understand his birds, a study of the writing of avian ethnologists is a must.

PAUL A. JOHNSGARD

Hybridization

In a group containing as many species as the waterfowl family Anatidae, and which have been kept and bred under captive conditions for so long a period, it should not be surprising that an unusually large number of hybrids have been produced. Doubtless many of these hybrids were never fully authenticated nor have details of their occurrence been uniformly preserved, but records of some 400 interspecies hybrid combinations have been published. The most complete published summaries of these hybrid occurrences are *Bird Hybrids,* by A. P. Gray, and a paper by the writer that appeared in *The Condor,* Vol. 62, 1960.

Since waterfowl tend to hybridize in an almost directly proportional frequency to their nearness of evolutionary relationship, the majority of known hybrids have been produced by species of the same genus. In captivity, different genera of the same tribe will sometimes also hybridize, but it is unusual to have members of different tribes hybridizing except under forced circumstances. An exception to this last statement might be necessary for the wood duck, however, since it has hybridized with at least 26 other species, of which only 3 have been perching ducks. This record of hybridization is exceeded among waterfowl only by the mallard, which has hybridized with at least 40 other species, even including certain species of geese.

Complete graphic summaries of all known interspecies hybrids among the Anatidae were published by me in the paper mentioned above, and thus for the most part need not be repeated. However, some additional records for the dabbling ducks and the pochards have since come to my attention, and certain minor changes in the summaries for these two groups have become necessary. Therefore, updated versions of these two diagrams are included here, and provide a means by which one can readily determine whether any intratribal hybrid combination has previously been reported and, if so, its relative frequency and the fertility of the hybrid combination when known. Species of these two groups that are never known to have hybridized are not included in the diagrams, although recently a record of the African black duck hybridizing in captivity with the yellow-billed duck has been published by W. R. Siegfried.

Waterfowl hybrids, although disadvantageous to the aviculturalist wanting to maintain the purity of his stock, are of such great biological interest to scientists that it is of special importance that records of their occurrence be made whenever they are produced, either accidentally or by design. It is of particular interest to know whether the hybrids survive to reproductive maturity, and if they attempt and succeed in reproducing, either among themselves or with one of their parental species. Hybrids occurring between species of markedly different adult plumages or reproductive behavior patterns are potentially valuable for the genetic study of these characteristics. Hybrids also may provide sources of materials for chromosomal analysis studies or for biochemical analysis, such as egg-white proteins. These sources of data may give additional clues as to the genetic diversity between the parental species that pro-

Waterfowl Hybridization

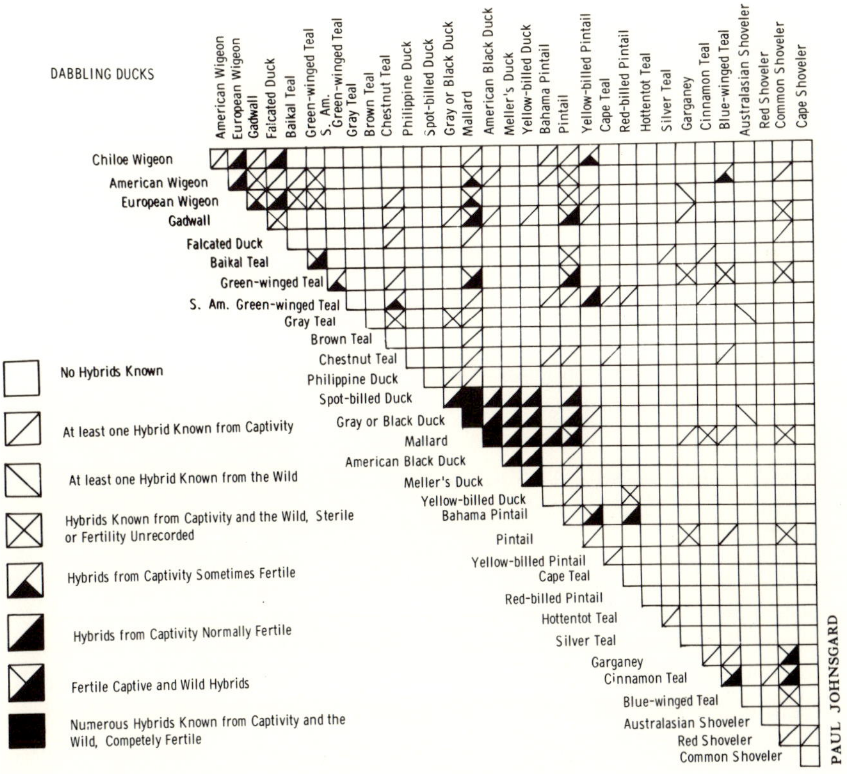

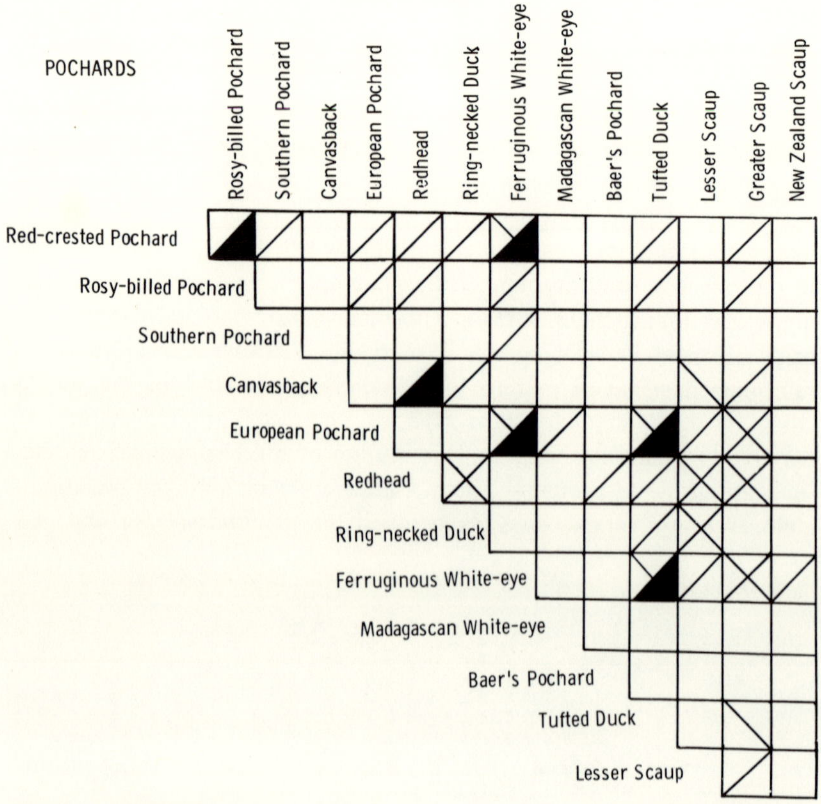

Waterfowl Hybridization PAUL JOHNSGARD

duced the hybrids. Undoubtedly many waterfowl and other hybrids are produced every year by aviculturalists that would be of interest to biologists for various reasons such as these, yet unfortunately they are often destroyed or no permanent record of their occurrence is kept. The author would be pleased to receive information relative to such hybrids, especially those combinations that would represent new records or provide additional information relative to the hybrid records included in the diagrams accompanying this article.

The remarkable degree of hybrid fertility which is found in many duck hybrids provides the aviculturalist with an unusual opportunity to establish experimental hybrid lines for various reasons. It would be interesting, for example, to see whether the fertility and reproductive vigor of the Hawaiian duck might be improved by producing mallard X Hawaiian duck hybrids and subsequently breeding the hybrids back to the Hawaiian duck. In this case, since the Hawaiian duck is obviously an island derivative of the mallard that has become greatly inbred, it seems unlikely that any serious damage would result from the introduction of some "fresh" mallard genes, and it perhaps might help to stimulate reproduction in the captive stock of this relatively rare and delicate species. When such crosses and subsequent back-crosses have been carried out previously, the physical traits of the species that is introduced into the bloodline seem to be fairly rapidly lost in individuals that have a parentage of one-fourth or less of that species. Undoubtedly, however, some of the introduced genes will persist in the population indefinitely, and periodically traits of the alien species are likely to appear in the progeny. Likewise, allowing the hybrids to breed *inter-se* normally results in a second generation of birds that range in appearance all the way from those nearly identical to one parental species through intermediates to ones closely approximating the other parental species. In general, the more closely related the two parental species, the greater the second-generation hybrid variability.

Some experimental work has been done with hybrids involving three or more species. The most extensive crossing of this type was done by J. L. Bonhote, during the early part of this century. He was able to produce hybrids that were derived from up to as many as five different sources, namely mallard, New Zealand gray duck, Meller's duck, spot-billed duck, and common pintail. He found that infertility in such hybrids tended to be reached by the third generation from the last cross with a pure species, but that breeding back to a pure species would restore hybrid fertility. With decreasing fertility, the hybrids showed an increasing tendency toward paleness and albinism of the plumage, which would suggest that birds such as laysan teal and other small, inland-dwelling ducks that tend toward albinism also suffer from a genetic deterioration that might be improved by temporarily breeding back to a vigorous species that most nearly approximates their ancestral origin.

Recent research on various duck species has shown that the greatest likelihood of producing "sexual imprinting" between species comes from rearing individuals of two species together as nest mates, and that male ducks thus reared are more prone to become sexually attracted to females of other species when reared with them than are females when reared under the same conditions. Thus, for persons wishing to prevent the production of hybrids, the rearing together of mixed broods representing different species should be avoided. The use of foster parents of a different species is therefore less likely to result in later sexual attachment and possible hybridization than is the use of foster nest mates.

The predominance of males among reported hybrids probably results from two reasons. First, male hybrids between most species are much more likely to be recognized as such than are females, owing to the usually considerable similarity among females of most species. Secondly, duck hybrids which are completely sterile often tend to assume male-like plumages, even those which are genetic females. It would seem that male hybrids more frequently are sexually active and more commonly are reported to be fertile than are female hybrids of the same cross. In several instances male hybrids have successfully been bred back to females of one of the parental species, but not with females of the same cross. In addition to possibly having reduced fertility, hybrids often seem to show a generally lowered level of pair-forming behavior and reduced ability to successfully compete with pure individuals for mates. At times this leads to the formation of homosexual pairs among the hybrids, judging from my personal experience with mallard X pintail hybrids.

In summary, waterfowl hybrids provide a subject of very considerable biological interest, and their occurrence and fertility are of special interest to geneticists and evolutionary zoologists. The crossing of very closely related species or geographic races may additionally provide some potential for establishing desirable traits in a captive line, but also produces the risk of incorporating undesirable genes into populations.

J. L. SKINNER

Infertility and Artificial Insemination

Pintails

Man has long been concerned with the reproduction of various kinds of bird and animal life in captivity. Herdsmen and curators have searched for means of assisting nature along reproductive lines. Various attempts at impregnation of female domestic animals were made as early as the 14th century, but not until Quinn and Burrows published their work in 1936 did a relatively simple but effective procedure for the insemination of fowl become widely known.

These researchers, working with chickens and turkeys, proved conclusively that artificial insemination could be made to work effectively in increasing the reproductive ability of individual birds or to make matings where physical differences in size, body conformation, or temperament would otherwise have prohibited any offspring.

Following the lead of these men, others became interested in the picture; and today we have artificial insemination widely used throughout the commercial turkey industry. It is also not infrequently used among commercial poultry breeders in the maintenance of inbred lines and other physically weak populations. Poultry research workers and breeders have used artificial insemination as a means of increasing the number of different matings that can be made with a very limited complement of equipment. Poultry fanciers who are interested in the protection of feathers and who in some cases have birds of unwieldy body conformations may use artificial insemination as a means of avoiding feather trimming, feather soil, or damage due to the action of normal mating. Other segments of the bird fancy have employed the techniques of artificial insemination in specific instances to overcome natural deficits in fertility brought on by advancing age of birds, unnatural habitat, and a variety of other conditions.

Johnson, and later Kinney and Burger, worked out procedures for the insemination of geese artificially, and various research workers have from time to time employed the technique with domestic as well as wild ducks.

The equipment needed for satisfactory artificial insemination is very simple. The very simplest of tools will give satisfactory results and are most frequently used when the numbers of birds being inseminated are small. More complicated equipment such as semen injection gun, semen collection aspirators, and temperature controlled containers for holding semen are sometimes employed for the large commercial operations, but for the fancier, the hobbiest, the collector, and the curator, minimal equipment is adequate and will give satisfactory results.

Artificial insemination is a procedure that can be most likened to an art rather than a science. Certain basic conditions are necessary for artificial insemination to be successfully employed. When these conditions are met, the success of artificial insemination depends to a considerable degree on the practice and patience of the inseminator. The procedures are not highly technical, but a basic knowledge and appreciation of the bird's anatomy is necessary.

The Male

For satisfactory results, the male(s) used for artificial mating should meet the following conditions:

1. Must be mature, in good health, and physically normal representatives of their variety.

2. Should be tame or at least not terrified when restrained or handled.

3. Must be sexually active. (This is an important consideration in those birds that are sexually active during a limited season.) Artificial light may be used to advance the season in some instances.

4. Should be free from external parasites. Some parasites irritate the vent area and make exposing of the male organ both difficult and painful to the bird.

5. Should be kept apart from, but preferably in sight of, the females.

6. Should not be subjected to extremely high temperatures or allowed to become overheated.

The male is held with his head toward the operator and keel lying in the palm of the left hand. The right leg is secured between the thumb and first finger. The left leg is secured between the first and second fingers in the case of small varieties. With larger birds, the left leg is more comfortably held between the second and third fingers. The back is stroked from midpoint toward the tail by the operator with his right hand, while the fingers of the left hand massage the abdomen from below. After several vigorous strokes, the right hand is transferred from the back to a position where pressure can be applied by the thumb and forefinger to either side of the vent. Simultaneously, pressure is applied to the abdomen by fingers of the left hand. This will normally extend the copulatory organ and a flow of whitish liquid will appear on it. This is the semen and should be caught by an assistant in an eye cup or any other small smooth-edged vessel. In some instances, the copulatory organ may not be completely extended. Semen collection may still be possible as it can often be seen to flow over the surface of the partially everted vent.

Males should be stimulated and semen collection attempted *immediately* after catching. To hold a male (even a relatively tame one) for only a few moments may interfere with your ability to obtain semen from him.

Successful semen collection is usually the result of an experienced operator and an experienced subject. First attempts at "working" inexperienced males often produce disappointing results. Some males will pass feces at the same time as the semen is discharged. Every effort should be made to collect only the semen. Feces are generally rather high in uric acid and semen requires a neutral-to-alkaline medium. Therefore semen contaminated with fecal material will usually produce poor if any results. Withholding water and feed for four to six hours prior to collection greatly lessens the chances of the semen being contaminated by an unexpected fecal discharge.

The volume of semen discharged varies considerably from one individual to another. Most males will produce between .1cc and .4cc at each successful collection.

Individual males vary considerably in the time needed to replenish their semen supply but normally collection once every two to four days is possible without harm to the bird.

If at all possible, semen should be used as soon as possible after collection. It may be held for brief periods of time (one or two hours) with but little loss in fertilizing capacity and for longer periods under carefully controlled conditions. A few suggestions for holding semen include prevention of dehydration (drying) and keeping the semen temperature below the body temperature of the male that produced it.

The Female

For satisfactory results, the female(s) used in artificial insemination should meet the following conditions:

1. She *must* be in production. Attempts to expose a female that is not in production will result in injury.

2. There should *not* be a hard-shelled egg discernible in the lower portion of the oviduct.

3. Pressure on the female's body should be relaxed as soon as possible after insemination. She should be released completely as soon as convenient. In the process of releasing pressure and the return of the oviduct to its normal position, the semen is drawn inward.

Throughout the process of handling and exposing the female, the operator should bear in mind that she is delicate and gentle treatment is a must.

The hen is held and stimulated in much the same way as the male. When pressure is applied following the preliminary stroking and massage, the vent will evert and an orifice appear on the left side. The orifice may appear as a round rosette or as a cleft or skin overfold. This orifice is the terminal opening of the oviduct. The semen is placed into this opening to a depth of ¼ to 1 inch with a 1cc syringe, a medicine dropper, or similar device.

In some instances, a second orifice may be observed on the right side of the vent. This second oviduct is rarely functional but if it appears about the same size as the one on the left, insemination in it is recommended. Some of these right oviducts have been observed to be functional. Although this condition is rather rare it is reported to occur occasionally in ducks and may be more frequent in inbred populations.

When making individual matings (1 male with 1 female) the entire collection of semen is normally used. However, various studies have shown that good fertility can be achieved with as little as .05cc of semen per insemination.

The frequency of insemination needed to produce a satisfactory fertility will vary somewhat from one female to another. Generally, one insemination per week is adequate. It may be advisable to inseminate at more frequent intervals at the onset of production but once fertility has been established, insemination at weekly intervals is sufficient to maintain fertility at a satisfactory level. Actual fertilization (the uniting of the sperm and ovum) usually takes place in the funnel or infundibulum area of the oviduct. This happens shortly after the yolk has separated from the ovary and before the albumin has been placed around it. Since ovulation frequently occurs within minutes after egg laying and because most of the eggs are laid early in the day, insemination should be done in the latter part of the day. If no nearly formed egg is present in the oviduct, the sperm can more easily move to the infundibulum area. Fertile eggs will normally be had from 48 to 96 hours following insemination and may occur for as long as 3 weeks thereafter. The percentage of fertile eggs from a flock will begin to drop in 5 to 7 days and be unsatisfactory beyond 10 days. Turkeys generally have longer intervals of fertility than some other birds, and geese show considerable variation from hen to hen.

There can be many reasons why eggs fail to hatch. When eggs are determined to be infertile (as opposed to early embryonic deaths), artificial insemination can frequently be used to beget another generation. In cases where failure to mate is caused by shyness, physical limitations, lack of purposeful interest, or social incompatibility, artificial insemination is a useful tool. Artificial mating, however, cannot overcome poor management practices, genetic lethals and/or differences, and health of the birds.

The inexperienced person should be warned against drawing hasty conclusions. Artificial insemination of many kinds of birds is a relatively simple procedure but does require practice. Wild bird and waterfowl breeders would do well to gain initial experience with some type of common poultry. Cornish bantams are excellent for this purpose.

Bibliography

Burrows, W. H. & J. P. Quinn. The Collection of Spermatozoa from the Domestic Fowl and Turkey. P.S. 15:19–24.

Burrows, W. H. & J. P. Quinn. Effective Dosages of Undiluted Semen in Artificial Insemination of Chickens. P.S. 17:131–135.

Burrows, W. H. & J. P. Quinn. A Method of Obtaining Spermatozoa from the Domestic Fowl. P.S. 14:251–3.

Buss, E. G. Artificial Breeding of Turkeys. N.T.F. sponsored survey at Pennsylvania State College.

Ferebee, D. C. Insemination Procedures. California Extension Service "Fowl Facts" April '65.

Frank, A. H. Artificial Insemination in Livestock Breeding. U.S.D.A. Circular 567.

Johnson, A. S. Artificial Insemination and Duration of Fertility in Geese. P.S. 31:No. 3

Kinney, T. and R. E. Burger. A Technique for the Insemination of Geese. P.S. 39: No. 1.

Quinn, J. P. & W. H. Burrows. Artificial Insemination of Fowls. Journal of Heredity 27:169–173.

Van Wambeke, F. Preliminary Results with New Diluents. F. Reprod. Fert. XIII (571–575).

Wiegers, H. L. Artificial Insemination of Turkeys. University of Nebraska Circular 96.

PART VI

Nutritional Requirements

EARL R. HOLM and M. L. SCOTT

Dietary Needs

Pintails

The nutritional requirements have been well established for domestic animals and poultry because of their great economic value as food for the table. On the other hand, little has been known of the dietary needs of wild waterfowl since they have no economic value for food. An experimental program was undertaken in New York by the Conservation Department to attempt to extend the breeding range of certain species of dabblers and divers through artificial stocking (Holm, Bauer, 1959). This program was predicated on the development of mass production techniques as well as determining the nutritional needs of the various candidate species. Since there was such a great variation in the food habits of the surface-feeding ducks in contrast to the divers, there were certain basic questions that had to be resolved. The principal questions involved proteins. How much protein was required in the diet of a duck at various ages? How much of this must be animal protein? A review of the literature covering the habits of wild ducks was a starting point.

A number of surveys have been conducted (Martin and Uhler, 1939; Girard, 1941; Stoudt, 1944; Mendall, 1949; and Stollberg, 1950) on food habits of wild ducks in different parts of the country as determined by analyses of stomach contents. While these studies yield information of value to waterfowl management, they do not furnish information on quantitative nutritional requirements.

Analysis of stomach contents from black ducks in the state of Maine by Mendall showed that there is considerable variation in the relative amounts of vegetable and animal foods consumed at different localities in the state, even at the same time of the year. In the fall, approximately 85 percent of the food was of vegetable origin and 15 percent was of animal origin for specimens from inland freshwater areas. In specimens from inland tidewater areas, the vegetable foods amounted to approximately 90 percent and the animal foods to about 10 percent. On the other hand, the food of black ducks from coastal water areas consisted of approximately 75 percent animal material and only 25 percent vegetable material. In the winter, the amount of animal food consumed by ducks from the coastal waters amounted to 87 percent of the total diet. Examination of the stomach contents of downy young black ducks from inland freshwater areas showed that about two-thirds of their entire diet was made up of animal foods. Insects, principally larvae, were predominant. Snails, freshwater clams, and mussels were also found in large amounts in the diets of the young ducklings. Results of the stomach examinations, together with field observations by Mendall and associates, showed that after the ducklings were a week to 10 days old their intake of vegetable material gradually increased until by the time they were half-grown their food habits were similar to those of adult ducks.

It is apparent from the studies referred to that the animal protein intake of even a single species is quite variable. When one considers the differences in feeding habits between dabbling ducks and diving ducks it must be considered probably that even greater variations in animal versus vegetable food intakes exist among the various species of wild waterfowl. The fact that the intakes of widely different types of foods may vary considerably, depending solely upon the types of foods available in given feeding areas, indicates that the relative proportions of the various types of foods eaten may not be of primary importance. It appears that wild waterfowl are able to balance their diets with widely different kinds of foodstuffs, the primary considerations being the availability of sufficient protein, carbohydrates, fats, minerals, and vitamins regardless of the sources of these fundamental nutrients.

Studies of domestic fowl show that the protein requirements fall into definite classes. Chicks require only 20 percent. For young turkeys, the requirement is in the neighborhood of 28 percent (Scott, Heuser and Norris, 1948). On the other hand, domestic ducklings maintain satisfactory growth and development on diets containing as little as 15 to 17 percent protein (Scott and Heuser, 1951). In all cases, best results are obtained when the diet contains some animal protein, such as fish meal, but in no instance is the amount of animal food needed in excess of 10 percent of the diet.

With this background, studies were undertaken to determine, first, the approximate requirement of total protein and, secondly, the approximate amount of animal protein required by wild ducklings (Holm, Scott 1954). Since high protein commercial game-bird rations containing about 28 percent protein were most commonly used for starting ducklings, one of these was compared with an experimental ration containing approximately 19 percent protein which was known to promote excellent growth in domestic ducklings. Both rations were fed in pelleted form together with scratch grains on a free-choice basis. The data showed that the experimental diet promoted better growth and better feed efficiency than the commercial game-bird ration and indicated that the nutritional requirements of young dabbling ducklings are more satisfactorily met by the feeding of a 19 percent protein diet together with scratch grains on a free choice basis than through the use of a diet higher in protein content. This indicated that the protein requirement of wild ducklings approximates that of domestic ducklings.

The next study was conducted to determine the optimum amount of animal protein required by both dabbling and diving ducks. Mallards, pintails, and redheads were used in the first series, and later canvasback, scaup, and ringneck ducks were used. The different species were divided into three groups. Group one received a diet containing 8 percent fish meal; group two received 4 percent fish meal; and group three received a diet containing 16 percent fish meal. The levels of all other nutrients in these diets were maintained as nearly equal as possible.

The ducklings were weighed weekly until 21 days of age. For all species the diet containing 8 percent fish meal as the sole source of animal protein promoted the best growth, although the results with the other diets were not markedly inferior. This indicated that the exact amount of animal protein in the diet is not critical when proper measures are taken to supply balanced amounts of protein, carbohydrates, fats, minerals, and vitamins in approximately the same amount that have been found to be required by domestic ducklings. It was of particular interest that equivalent results were obtained with both dabbling ducks and diving ducks although their natural feeding habits differ widely. Upon the completion of these formal controlled studies, practical rations were developed and modified over the years which proved to be adequate for all species handled. These did not include such species as the mergansers and the sea ducks, which feed exclusively on fish. The composition of the starter, grower, and breeder diets are given in Tables 1, 2 and 3.

Each of these rations have the built-in protection, in the form of certain nutrients, to meet the specific requirements of waterfowl. For example, waterfowl are prone to develop leg weaknesses or enlarged hocks. This condition also occurs in turkeys. Increased levels of niacin will correct this condition. These leg disorders are likely to occur when chicken or broiler feeds are used and they are quite common when dog meal is provided as part of the diet.

Standard commercial duck rations are not generally recommended for wild waterfowl since they are designed for maximum growth and the development of a carcass to be processed for food. Additives are frequently included as growth stimulants and others to facilitate removal of feathers or to add extra profit for the feathers which are sold as a by-product. Protamone or iodinated casein is one of these. Such a diet has no place in an operation of producing wild waterfowl for the development of a superior breeding stock. Dog meal or dog biscuit is not recommended for similar reasons. These feeds are scientifically formulated for dogs, not waterfowl. In most instances, the fat content in dog foods is much higher than waterfowl can tolerate, particularly when mixed grains are fed as a part of the regular program. Most ducks have a tendency to have excessive fat deposits in the body cavity, as well as around the heart and gizzard. The liver stores some of this fat as glycogen. But fatty livers often result in fatality among waterfowl. Adequate levels of choline chloride afford protection against fatty liver or fatty infiltration of the liver. If the level of this nutrient is not adequate in the diet, it must be supplemented.

Controlled studies have shown that the growth rate of birds is dependent on the ratio of the percent of protein to the metabolizable energy. This latter is the amount of carbohydrate (energy) in the diet

Table 1

Wild Duck Starter Ration

INGREDIENTS	LBS. PER TON
Corn meal, No. 2, yellow	933
Oats, heavy, pulverized	200
Wheat standard middlings	300
Soybean oil meal, 50% protein, low fiber	250
Fish meal, 60% protein	160
Fish solubles, dried basis	10
Dried brewers yeast, 40% protein	20
Whey, dried product, 55% lactose	20
Alfalfa meal, dehydrated, 17% protein (100,000 A/lb.)	60
Dicalcium phosphate	10
Calcium carbonate, ground	30
Salt, iodized	5
Manganese sulfate, feed grade	0.5
Copper sulfate	0.5
Zinc carbonate	0.25
DL-methionine (Hydroxyanalog)	1

VITAMINS	AMOUNTS ADDED PER TON
Stabilized vitamin A, USP units	10,000,000
Vitamin D_3, ICU	1,500,000
Vitamin E, IU	5,000
Riboflavin, grams	3
Choline chloride, grams	112
Niacin, pure, grams	40
Calcium pantothenate, pure D-isomer, grams	6
Vitamin K (menadione sodium bisulfite), grams	4
Vitamin B_{12}, milligrams	6
TOTAL	2,000 lbs.

CALCULATED ANALYSIS:	
Protein, %	20.0
Fat, %	6.5
Fiber, %	4.0
Calcium, %	1.2
Phosphorus, %	0.7

Starter ration to be fed for first three weeks with insoluble grit available at all times.

Table 2

Wild Duck Grower Ration

INGREDIENTS	LBS. PER TON
Corn meal, No. 2, yellow	753
Oats, heavy, pulverized	400
Wheat standard middlings	300
Soybean oil meal, 50% protein, low fiber	280
Fish solubles, dried basis	10
Fish meal, 60% protein	100
Brewers dried yeast	20
Whey, dried product, 55% lactose	20
Alfalfa meal, dehydrated, 17% protein (100,000 A/lb.)	80
Dicalcium phosphate	10
Calcium carbonate, ground	20
Salt, iodized	5
Manganese sulfate, feed grade	0.5
Zinc carbonate	0.25
DL-methionine (hydroxyanalog)	1
Copper sulfate	0.5
Santoquin	0.25

VITAMINS	AMOUNTS ADDED PER TON
Stabilized vitamin A, USP units	12,000,000
Vitamin D_3, ICU	1,500,000
Vitamin E, IU	5,000
Riboflavin, grams	4
Choline chloride, grams	250
Niacin, grams	40
Calcium pantothenate, pure D-isomer, grams	10
Vitamin K (menadione sodium bisulfite), grams	4
Vitamin B_{12}, milligrams	6
TOTAL	2,000 lbs.

CALCULATED ANALYSIS:	
Protein, %	19.0
Fat, %	5.0
Fiber, %	4.5
Calcium, %	1.0
Phosphorus, %	0.7

Grower ration to be fed from 21 days of age to maturity with scratch grains free choice.
Grit to be available at all times.
Grower plus scratch grain free choice serves as maintenance ration and for nonlaying breeders.

Table 3

Duck Breeder Ration

INGREDIENTS	LBS. PER TON
Corn meal, No. 2, yellow	905
Oats, heavy pulverized	200
Wheat standard middlings	100
Fat, stabilized grease or vegetable oil	60
Soybean oil meal, 50% protein, low fiber	250
Fish meal, 60% protein	150
Brewers dried yeast, 40% protein	50
Corn distillers solubles, 25% protein	50
Whey, dried product	50
Alfalfa meal, dehydrated, 17% protein (100,000 A/lb.)	50
Dicalciun phosphate	20
Calcium carbonate, ground	100
Manganese sulfate, feed grade	0.50
Santoquin	0.25
Zinc carbonate	0.25
Salt, iodized	10
DL-Methionine (or hydroxyanalog)	4
Copper sulfate	0.5

VITAMINS	AMOUNTS ADDED PER TON
Stabilized vitamin A, USP units	8,000,000
D-activated animal sterol, ICU	1,000,000
Vitamin E supplement, IU	20,000
Riboflavin, grams	3
Calcium pantothenate, pure D-isomer, grams	10
Vitamin K (menadione sodium bisulfite), grams	2
Zinc bacitracin, grams	50
Vitamin B_{12}, milligrams	6
TOTAL	2,000 lbs.

CALCULATED ANALYSIS:		AMINO ACIDS	% OF THE PROTEIN
Protein, %	19.4	Arginine	6.9
Fat, %	6.7	Lysine	6.2
Fiber, %	3.8	Methionine	3.24
Calcium, %	2.7	Cystine	1.8
Phosphorus, %	0.77	Tryptophan	1.3
Metabolizable energy,		Isoleucine	5.75
Cal/lb	1320	Valine	5.6
ME/P	68		

Breeder ration to be fed when laying commences and until egg production ceases. Grit and shell to be available at all times.

that is actually utilized by the body. When both the protein and the energy levels are increased in a diet, the rate of growth will be accelerated. It is this principle which permits the Long Island domestic duck growers to market ducks weighing 8 pounds and over at 8 weeks of age. The same principle can be applied to wild waterfowl.

If the protein is adequate, a group of ducks fed a high protein level will finish off at the same weight as another group which is fed at a lower level. But it will take the latter group a little longer to reach the point where the weight gains level off. Thus there is no advantage to the producer of wild waterfowl to use a ration which stimulates rapid growth. Rather, it can be assumed that a ration which produces normal growth for the species would be the most desirable and it would be cheaper.

Many experienced growers are convinced that wild waterfowl must have a high protein diet, and one pelleted ration is fed to all of the waterfowl in the collection. This is in spite of the fact that many, like some geese, are grazers. Most of the lush grasses seldom exceed 17 percent protein. The young like insects soon after they hatch but in a short time their diet is similar to the adults. In many cases excessive protein can be detrimental and in some instances is toxic. The kidneys are the organs that must dispose of the excess protein in the form of urates. Additional water must be consumed to constantly flush the kidneys. If founts are allowed to go dry, kidney damage and death can result in a relatively short time. If water is withheld, this may occur even with optimum protein levels.

When ducklings eat litter or when gullets or gizzards are impacted by grass or other extraneous material, death usually results. But it is not starvation that is the direct cause. On close examination, one will find urate crystals in the kidneys which result from lack of water due to the impaction.

Gizzard erosion frequently causes mortality in waterfowl flocks. The addition of copper sulphate or oxide will correct this.

Zinc is likewise essential in the diet and is usually supplemented in the form of the oxide or carbonate. Zinc deficiencies cause poor egg production, impaired hatchability, and abnormalities in the embryo.

In view of the foregoing, the most practical method of providing a diet to meet the needs of wild waterfowl is to use a feed specifically designed for them such as the rations described in the tables. In many cases, it may not be possible to have these formulated and mixed in small quantities. On the basis of the reasoning set forth, neither chicken, broiler, nor commercial duck diets are to be recommended. There are, however, other specialty rations that are readily available. These are the turkey feeds.

On the basis of published studies, the basic nutritional requirements of game birds lie between the level of the chicken and the turkey. Thus, a good turkey ration will serve as an adequate game bird ration. Therefore, the turkey diets would be the feeds of choice for waterfowl if the specialized diets are not available. However, it has been pointed out that waterfowl do not require the high protein levels of turkey and game-bird feeds. Therefore, it would be more desirable to use a turkey grower with 20 percent protein as a wild duck starter. This could be supplemented with scratch grain between the second and third week. After the ducklings are 3 weeks old, the grower and grain should be provided in separate feeders, free choice.

After the birds are mature, they should be fed a maintenance diet. A combination of 10 percent grower pellets and 90 percent mixed grains is adequate until the laying season is about to begin.

During the fall and winter, the nutritional requirements provide only for the maintenance of body weight. Energy producing feeds, like scratch grain with corn or sorghums, are required to maintain body heat. As the temperatures fall lower in the northern latitudes, the intake of the energy producing feeds will increase. Thus, when the thermometer reading is below zero, the ducks and geese will consume more corn.

Practical Feeding Procedures

Feeding habits of the various groups of ducks and geese differ widely. Certain of the ducks feed entirely or almost exclusively on fish and other marine animal life such as clams, mussels, shrimp, and other crustaceans. These ducks include the mergansers, scoters, and eiders as well as numerous others. Thus, at the extremes there are the grazing geese which are strictly herbivorous, the groups of divers which are carnivorous—with other species between which are omnivorous. This poses some problems in setting up a practical feeding program.

Based on experience, it can be assumed that the diets described will meet the nutritive requirements of the surface-feeding duck and most of the common ducks. When ducklings reach the age of 3 weeks, scratch grains such as corn, wheat, barley, oats, and milo can be supplemented, which in effect serve to lower the level of protein as the ducklings increase in age. When pellets and grain are fed free choice, the ducks will be consuming approximately equal amounts about the time they are able to fly. For ducks that are pinioned or clipped, it will coincide approximately with the time the quills of the primaries become hardened. However, the fish-eating ducks require different treatment.

Many of these ducks that feed entirely on marine animal life are extremely difficult to rear. Some have been successfully started from day-olds by hand feeding small minnows and/or crayfish. But, with the development of trout foods in pelleted form, many of the problems have been resolved.

The pelleted trout feeds are frequently used for starting day-old ducklings of the fish-eating species. Some of these trout feeds have a protein level of about 45 percent. If a specialized feed meets the nutritional requirements of trout and other cold water fish, it seems reasonable to assume that it would meet the requirements of fish-eating ducks. In some cases trout and duck diets are fed in combination.

Those species of ducks that do not feed exclusively on marine animal life might well be fed a combination that contains a high level of the trout food. As they increase in age the trout food should be decreased, with a corresponding increase in the proportion of the duck diet. Observations of the droppings often serve as a rough guide in feeding.

There is a wide variation in the color and consistency of the droppings of waterfowl. Again the extremes are found in the grazing geese at one end whose droppings are green and semi-solid. In contrast the waste from the fish-eating cormorant is a splash of whitewash. Accordingly, the type and form of the feed that is provided to the various species should be compatible with their specialized digestive system.

Feeding the Breeders

Most feeding instructions provided by commercial feed manufacturers recommend providing the breeder diet a month before the laying season begins. But, these recommendations lack scientific support. It was pointed out previously that a maintenance ration should supply enough protein for tissue replacement and enough energy (carbohydrate) to serve as fuel to maintain the body weight. To meet the requirements of a laying breeder, additional nutrients must be added.

In developing a laying ration, such items as extra protein must be added to meet the needs of the egg. Extra calcium is required to form the shell. Extra vitamins and minerals (trace elements) are necessary to assure a high level of fertility, hatchability, and livability. If a duck is not producing eggs, the surplus nutrients are voided. There is no space in the carcass of waterfowl for extended storage of nutrients, except fat. There is no need for such extended storage when a balanced diet is provided that contains all of the essential nutrients that are consumed daily.

Furthermore, it has been demonstrated that a laying female, on a maintenance diet, will draw from her body reserves to produce normal eggs. Calcium, as well as the phosphorus required for the shell and development of the embryo, are drawn from the bones. If this is prolonged, the bones of the legs become softened to the extent that the female can no longer stand and support her weight. This condition occurs sometimes before soft-shelled eggs are laid.

Even though the diet is balanced for the layers, oyster shell or some other suitable form of extra calcium should be provided, free choice. Apparently there is a difference in calcium metabolism among different species and within the same species, making it desirable to furnish an extra supply.

The Importance of Grit

The gizzard serves as the "jaw" for waterfowl, and natural grit represents the "teeth." Many of the softer foods can be macerated by the gizzard without the presence of grit. Pelleted feeds, for example, do not require grit. On the other hand, it has been demonstrated that the gizzard is reduced in size and it becomes quite flaccid. But, by providing grit, the muscle is exercised and strengthened so it is comparable to the organ in its wild counterpart. Only hard, insoluble granite grit should be provided, since limestone or other calcium carbonate grit will upset the mineral balance of the ration. Grit is graded in size to meet the needs of growing waterfowl and adults. Chick size, intermediate, and hen size are most readily available. It should be provided when the ducklings are eating well at about 3 days of age. It is advisable to sprinkle it over the top of the feed at first, or it can be provided in a separate dish or feeders. Some operators incorporate 4 percent grit with the pellets or crumbles.

The Future in Nutrition Research

Though the feeds and feeding techniques for many species of waterfowl are largely improvisations, it appears evident that a scientific breakthrough is imminent. Sophisticated electronic equipment is available now and being used to determine the total protein level of specific organs, tissue, and feathers of any given species at any given age. Thus, it is possible not only to determine the percent of protein in the whole bird but also to analyze the percent of each of the amino acids which make up the total protein of the carcass. By a determination of the carcass composition of any species at any age, a diet may be formulated which will provide the proper nutrients in the proper proportions. The basic technique for preparing the material for analysis is quite simple.

At the end of the starting period for ducklings (3–4 weeks), representative birds are killed, the feathers removed, and the carcasses are ground separately for each bird. Each sample is then freeze-dried and thoroughly mixed. Further samples can be taken at 4-week or other intervals and treated in the same manner.

Protein determinations are conducted on samples representing the whole birds (carcass and feathers) and also on the whole carcass minus the feathers. The amino acids are determined on the carcass and are recorded as the percent of carcass protein. Fat and ash are usually determined only as percentages of the whole bird.

It would be expected that domestic birds and animals would receive highest priorities for analysis of this type, again, because of their economic importance for food. On the other hand, it augurs well for speedy determination of the nutritional requirements for any of the rare and endangered species which may be propagated artificially as a last resort. At least it provides new hope for the near future.

Bibliography

Girard, G. L. 1941. The Mallard: Its Management in Western Montana. Journ. Wildl. Mgt. 5(3):233–259.

Holm, E. R. and Bauer, R. V. Potentialities of certain Species of Waterfowl for Mass Production. New York Fish and Game Journal 6(1):1–45.

Holm, E. R. and Scott, M. L. Studies on the Nutrition of Wild Waterfowl. New York Fish and Game Journal 1(2):171–187.

Martin, A. C. and F. M. Uhler. 1939. Food of game ducks in the United States and Canada. U.S. Dept. of Agr. Bull. 634.

Mendall, H. L. 1949. Food habits in relation to black duck management in Maine. Journ. Wildl. Mgt. 13(1):64–101.

Scott, M. L. 1951. The grass juice factor in turkey nutrition. Poultry Sci. 30:293–297.

Scott, M. L. and G. F. Heuser, 1951. Studies in duck nutrition. 2. Studies of protein and unidentified vitamin requirements. Poultry Sci. 30:161–167.

Scott, M. L., G. F. Heuser and L. C. Norris. 1948. Energy, protein and unidentified vitamins in poult nutrition. Poultry Sci. 27:773–780.

Scott, M. L. and L. S. Jensen. 1952. The effect of antibiotics upon the requirement of turkeys for unidentified vitamins. Poultry Sci. 31:986–993.

Stollberg, B. P. 1950. Food habits of shoal-water ducks on Horicon March, Wisconsin. Journ. Wildl. Mgt. 14(2):214–217.

Stoudt, J. H. 1944. Food preferences of mallards on the Chippewa National Forest, Minnesota. Journ. Wildl. Mgt. 8(2):100–112.

WILLIAM HOLLENBECK

Meal Worms

Meal worms are easy to raise, but you must start 6 months ahead of time. They should be used only as a diet supplement.

Go to your local feed mill and get 50 pounds of wheat bran and while you are there get some meal worms. Look around in old burlap-bag piles or paper feed bags.

Get some large tin cans with tops and punch a few holes in the top for air. Fill one-half to three-quarters with wheat bran and a dozen meal worms. Add a potato every month for moisture.

Your meal worms will go into a chrysalis and hatch as a black beetle. The beetles will breed and lay eggs which hatch into very small meal worms.

When you look at the wheat bran and you see the flakes move, take a handful and add to other cans as they are too small yet to see. They then grow, shedding their skin several times. You can use them any size you want. Small or baby ducks see them wiggle and crawl when you drop them in front of them and many times this gets a duck to start eating.

Gradually your wheat bran will be eaten up and go into a powder. Just add fresh bran and you will have meal worms the year around.

In the north they go dormant in freezing weather but awake in spring and you will have a good supply just in time.

CLIVE ROOTS

Maggots and Meal Worms

We regularly read about the problems arising from feeding maggots and meal worms to soft-billed birds—problems which occur when these larvae form the sole items of diet of nestlings, at least for the first week or so; and of bee-eaters and other exacting hawking species which are loathe to accept inanimate foods. Almost as frequently we hear of attempts to "improve" the quality of both maggots and meal worms by the addition of certain items to their diet or by dusting them with a supplement before they are fed, without really knowing what food components are lacking and need to be increased.

In an effort to throw some light upon this very important aspect of aviculture, rather than rely upon assumption, samples of maggots and meal worms have been submitted for analysis, and the results have proved very interesting to say the least. In order to fully appreciate the notes which follow, the results of the analyses will be given first. Only freshly killed insects were analysed.

	Blowfly Maggots	*Meal worms*
Fat	8.4 %	12.0 %
Protein	16.2 %	20.8 %
Carbohydrate	0.89%	1.86%
Calories per 100 grams	150	204
Fibre	2.2 %	2.0 %
Moisture	70.6 %	61.9 %
Calcium	0.02%	0.03%
Phosphorus	0.2 %	0.27%

From these analyses it will be seen that the *calcium and phosphorus* content of both forms of larvae is minute, and is *certainly insufficient to rear young birds*. It is known that an adult bird can mobilize 25 percent of the calcium required for egg-shell deposition from its body, and while it would probably survive for a number of years on a diet so low in calcium and phosphorus, it is doubtful if such an intake would contribute sufficient for the bird to be able to produce eggs. We know that softbills seldom become egg-bound, yet there have been two instances recently at the Winged World when a Thailand hoopoe became egg-bound after laying two perfect eggs in each clutch. All other conditions were perfect and were not thought to have contributed to that ailment. Due to the fact that these birds were in a community aviary where maggots and meal worms were available all day for the more exacting species, the hoopoes had concentrated their attentions upon these and had ignored the insectile mixture which would have provided a balanced diet. Needless to say this pair have now been moved to quarters where they can receive a more balanced diet, and the indications are that we may be more successful in the future.

The relative lack of calcium and phosphorus in these larvae could have been foreseen anyway, of course, as they do not possess the skeletal system which contains the majority of the ash content of the vertebrate body. The skeletal deformities and other symptoms which nestling insectivorous birds often show in captivity are certainly due to the lack of calcium and phosphorus in their diet, and possibly to the fact that the small quantities of each are present in the wrong ratio, as much more calcium than phosphorus is required to build bones. The vitamin requirements, where the formation of bones is concerned, will be discussed later, as will the means of administering supplements to purely insect feeders.

The deficiency symptoms shown by nestlings which were reared or almost reared upon these larvae alone have mainly been associated with the feet or legs, although perhaps it would be truer to say that it was most obvious there, as they were twisted, weak, and useless. Beaks, too, have failed to harden and take their correct shape.

It will also be seen from the analyses that the fat content of meal worms is higher than that of maggots. The same applies to the protein and carbohydrate content. Rather than compare the content of each of these food nutrients separately, but as a total calorific content, it is necessary to understand a little about the calorie values of foodstuffs. This is an aspect of nutrition which to date seems to have been applied only to human, and not to bird diets.

There are approximately 4 calories per gram of protein, just under 4 in every gram of carbohydrate, and over double the quantity, just over 9 calories, in each gram of fat. When the composition of a diet is known it is a simple matter to multiply the percentage of protein in every 100 grams of food by 4, and so on. On this basis the calorie content of meal worms is 204 and maggots 150 per 100 grams (approximately 4 oz.).

While not being the ideal ratio of fat and carbohydrate, as birds can assimilate the latter easier, insectivorous species have survived for several years on maggots and meal worms alone. This life-span would no doubt be extended if vitamins and mineral salts could somehow be added to the diet.

Often we hear how "rich" meal worms are, and how sparingly they should be fed, the inference being that this applies to all birds. Where the insectivorous species are concerned the only thing rich about meal worms is the fantastic price they command. It is my contention that a diet of a normal insectile mixture which has been moistened with fats or oils, plus a few maggots daily, will provide a food intake far richer in fat, and therefore more likely to cause obesity, than a diet of meal worms alone.

Where frugivorous and omnivorous species are concerned it is a different matter, of course, as they definitely require a lower intake of calories than the insectivorous species. The addition of too many meal worms to a diet already containing fruits rich in carbohydrate and fatty insectile mixtures could just tip the scales in favor of obesity. Also, there is the over-stimulating effect which meal worms have upon many species, particularly among the omnivorous softbills.

The moisture content of these larvae is very high, being well over half the content of the meal worm and almost three-quarters of the blowfly maggot. It is not really surprising therefore that the bee-eaters, and other species which are not known to drink, can derive enough moisture from their insect diet, which without doubt they do in captivity.

Unfortunately, analysing maggots and meal worms for their vitamin content has not yet been undertaken, as this entails a far more costly operation. At this stage it seems reasonable to assume that a deficiency of vitamin D is also associated with the leg and feet deformities already mentioned. Even if there were sufficient calcium and phosphorus in the diet to form bones, this could not be done unless vitamin D was present to assist in the metabolization of the minerals. In adult birds vitamin D is manufactured by exposure of the skin to sunlight.

The vitamin which is definitely absent from both larvae, or at least present in the most minute quantities, is vitamin A. We have heard of cases of half-closed, watering eyes, followed by asthmatic breathing and eventual blindness and death, of bee-eaters fed for months solely upon maggots and meal worms; also of similar symptoms and losses among newly imported flycatchers, often so dependent upon live food for many weeks. Frequently I have heard it said that this is "due to something in the maggots, probably not cleaned properly." On the contrary, it is not something in the maggots, but something they are lacking. This has been proved conclusively where the exacting East African wattle-eyed and puff-backed flycatchers are concerned. A few which were imported on a diet of locusts and meal worms defied all attempts at meating off, and were restricted to a diet of maggots and meal worms as they would not even accept crickets and blowflies. Obviously they could not be released, which is the normal procedure in the field with newly trapped birds which will not accept inanimate foods. In a desperate attempt to save the surviving members of the group, a liberal dose of Abidec (rich in vitamins A and D) was administered direct into the beak and within two days the symptoms of watering eyes, dejection, and heavy breathing had cleared up. The mistake was then made of adding the water soluble Abidec to their drinking water, and soon afterward it became necessary to again administer drops directly into their beaks. It soon became obvious that these flycatchers, like the bee-eaters, did not drink, and we have since added the vitamin A to the live food. We now use the oil form of vitamin supplement, as this adheres to the larvae better, and since commencing this treatment the symptoms mentioned earlier, which are characteristic of vitamin A deficiency in other animals, have not recurred.

Several methods of administering supplements to birds which are feeding solely upon maggots and meal worms have been tried, from all accounts. Often though these appear to have been supplements which were intended to increase the protein content of the diet rather than the vitamin and mineral content. As we have seen from the analyses protein is present in sufficient quantity, and it is the minerals which need to be added, plus vitamin A. Adding all the other available vitamins can do no harm either. The administering of baby or invalid foods to the larvae, both internally and externally, is not really solving the problem, as the food items which are needed will be reduced to infinitesimal quantities due to the small amount of the foodstuff which is ingested anyway.

The answer lies in providing purely a vitamin and mineral supplement in a highly concentrated form, but once again administering this is not a simple matter. There appears to be no evidence that vitamins and minerals can be stored or retained in the body tissues of maggots and meal worms. Therefore if these larvae are fed upon highly enriched foods, the small amounts present in their gut when they are eaten is all that birds will receive. Maggots are unsuitable for this method anyway, as their foodstuffs will be in varying stages of fermentation or decomposition if the temperature at which they are kept is high enough to stimulate feeding. The maggots should be cleaned before they are fed to the birds, which of course defeats the object of the exercise. A supplement can certainly be added to the meal worms' food, but this must be in a highly concentrated, finely powdered form.

Maggots are more suitable than meal worms if it is intended to add the supplement externally to the larvae, as they remain alive for some time even when liberally coated with honey and powdered with the supplement. Meal worms, not being adapted for a semi-fluid existence, do not survive such treatment and are then not as readily accepted. Where vitamins are concerned the best method seems to be the addition of a few drops of highly concentrated vitamin oil, which adheres to the larvae, and does not dry out as rapidly as the water bound form. Minerals in very finely powdered form stick quite well to these moistened larvae.

Alternatively the larva can be injected with liquid calcium, e.g. Collocal-D, and liquid vitamins if a very fine needle is used. Naturally the amounts injected are minute, but as the additives will be in highly concentrated form this should be sufficient to rear healthy youngsters, and will certainly fill the requirements of adult non-breeding birds—in which case such a messy and tiresome task will have been well worthwhile.

DON FOLEY

Duckweed for Ducks

Lemna minor (lesser duckweed) is an interesting plant. It is one of our smallest, of course, being fully contained in a frond, with rootlet, of only ⅛ inch across, but not *the* smallest by any means. It is found in suitable locations all across the country, from Maine to Florida and from Washington to southern California—also all across southern Canada, with a big bulge up into the northern parts of Alberta, Saskatchewan, and Manitoba.

When the phrase "suitable locations" is mentioned, it is meant both physically and chemically. Quiet, shallow waters are best, especially if they are not acid and if they are shaded. Sometimes, in this case, duckweed blankets the whole surface.

Ducks eat it quite a bit, probably because of its abundance, but this is true of most food plants. It is probably taken for its own food value as well as along with the insects that are found in it. However, I have no doubt that it is every bit as nutritious as even the *Potamogetons,* except for their seeds or winter buds.

I'm pretty sure that duckweed can be cultivated, because that's what it does for itself, both by seeding (probably rather rare) or by vegetative process. As cold weather approaches, it forms minute bulblets which sink to the bottom and so are safe from freezing. When spring comes they rise to the surface and grow. Quite possibly fish fanciers, water lily culturists, or aquarium owners have kept it multiplying indoors over the winter, although it is also possible that the plant must go through a dormant stage. Light and aeration, of course, would be necessary if it were held over winter indoors.

Some of the other *Lemnas* (of which there are three besides *minor*) are not floaters, as *L. minor* is mostly, but also live submerged most of the time—*L. trisulca* and *valdiviana*. These are the star duckweeds that are known at Howland Island and Montezuma. There is also a smaller one, a free floater called *L. perpusilla,* more southerly in distribution, which comes north only to Nebraska, Illinois, New York, and Massachusetts.

Francis M. Uhler, Biologist, Section of Wetland Ecology, Bureau of Sports Fisheries and Wildlife, has this to say regarding *Lemna minor* as food for ducklings, juveniles, and mature ducks. "Until botanical taxonomists can agree on what constitutes the so-called *Lemna minor* (and no two seem to be able to agree on this subject) it is not feasible to comment intelligently on this nebulous species." Twenty years ago I thought that I knew what *Lemna minor* was. Now the latest "Monography of Lemnaceae" by Edwin Horace Daubs (University of Illinois Biological Monographs 34, 1965) tries to separate nine American species of *Lemna,* several of which are of the *L. minor* type and have overlapping ranges. This past summer, while working up in southwestern Manitoba I noted that even the star duckweed (*L. trisulca*), a usually distinctive species little used by ducks, has a late summer form that superficially resembles the old *Lemna minor.*

Dale E. Harrison and Ernest O. Beal of North Carolina State University discuss "The Lemnaceae (Duckweeds) of North Carolina" (Journal of the Elisha Mitchell Scientific Society, Volume 80, No. 1, pages 12–18, May 1964) and list three species of *Lemna* of the orbicular type—*L. gibba, L. minor, L. perpusilla,* and the quite similar *L. valdiviana.*

Lawrence E. Hicks, formerly of the Department of Botany, Ohio State University, discusses five *Lemnas* in "The Lemnaceae of Indiana" (The American Midland Naturalist, Volume 18, No. 5, pages 774–789, September 1937). These include *L. trisulca, L. minor, L. minima, L. perpusilla,* and *L. cyclostosa.* The last mentioned species apparently is synonymous with *L. valdiviana.*

The taxonomic discussions involve so many variable characters, and the nominal species are so indistinct, that a report which purports to discuss the ecology and nutritive values of these poorly differentiated species would be presumptuous at the present time. I have not made such studies.

We have used large quantities of *Lemna* of the *minor* and *valdiviana* types and two species of *Wolffia* (*W. papulifera* and *W. columbiana*) as a supplement (not a substitute) along with farinaceous foods for our captive ducks of many species at the Patuxent Wildlife Research Center. We have also used considerable quantities of several species of pond weeds (*Potamogeton* and *Najas*), water purslane (*Ludwigia palustris*), water starworts (*Callitriche stagnalis* and *C. heterophylla*), and common *Elodea* (*E. canadensis*). Practically all our ducks seem to greatly prefer the duckweeds of the *minor* and *valdiviana* types and the *Wolffias* to any of the other succulent foods that we have used. We use the duckweeds (fed in a floating condition) as a daily supplement for the ducklings and as heavy feedings once or twice a week for the juvenile and adult ducks. These birds have remained in an amazingly healthy, vigorous condition.

We harvest the duckweeds in dipper-shaped, fine-meshed wire strainers and large "lard" cans in nearby fertile, natural waters. The heaviest growths of *Lemna* of the *minor* type usually are found in hard carbonate waters that are high in nitrogen and phosphorus, and have a summer pH of about 8.

Ponds that receive the run-off from barnyards often have especially luxuriant growths of *Lemna minor*.

I doubt that it would be practicable to produce sufficient quantities on small, nursery ponds to permit frequent heavy harvests of duckweeds of any species for use as feed for large numbers of waterfowl.

GLEN SMART

Inducing Difficult Species to Begin Eating

Feeding is a natural phenomenon that aviculturists usually take for granted. In the wild, the parent birds gain the interest of newly hatched young by a variety of methods, some peculiar to single species. In the case of precocial young, that is, young that are capable of feeding themselves and of leaving the sanctuary of the nest within a few hours after hatching, the feeding response is often elicited by the parent bird offering tidbits of food, *i.e.,* dropping it, picking it up again, and so on until the youngster begins pecking at it himself. Movement of live food does much toward creating interest in the youngster. Whatever the method used to create this interest, once the chick gets the idea, it usually becomes independent of the parents in its search for food.

Fortunately, most species of young ducks feed readily and there is little cause for concern in getting them to begin feeding. There are, however, those few individuals that present problems. Generally speaking, the young of shelducks, dabblers, and pochards present the fewest problems. Some young perching ducks are occasional problems and will require some coaxing. It is the young of the so-called sea ducks (goldeneyes, mergansers, scoters, eiders, harlequins, and long-tailed ducks) that are most apt to present consistent problems.

Nature has provided newly hatched birds with a "built-in" food reserve that provides adequate nourishment for the first 36–48 hours. This food reserve, the yolk, is metabolized gradually so that neither the need nor desire to feed exists. It should, therefore, be stressed that the bird is in no danger if it does not begin feeding immediately after being placed in the brooder.

If the chick is not feeding at the end of 48 hours and is not showing an interest in his surroundings, it would then be wise to have alternate plans for initiating the feeding response. There are many methods for getting these difficult species to begin feeding. These usually require initiative and patience, and the desired results may not be immediate.

I can relate several personal experiences and will suggest several others that have been successful for various people. This is not intended as a complete treatise on the subject as everyone has his own personal method that is most successful for him, and I suspect that a survey of aviculturists would result in an equal number of "most successful methods." Personal initiative and imagination are foremost in importance in the development of such methods.

Motion often plays an important part in these techniques. For instance, while I was affiliated with the Round Lake Waterfowl Research Station, Round Lake, Minnesota, we employed the use of a motorized feeder that had a small electric motor with a drive rod attached. This metal drive rod, with cross-pieces soldered to it, moved slowly up and down the shallow feed pan, causing the bits of food to move. The moving feed attracted the ducklings and, perhaps out of curiosity, they began to peck at it. Enough food would be obtained by this pecking activity so that the young would be feeding normally within a few minutes.

During the first few weeks of life, most waterfowl rely very heavily upon aquatic invertebrates for their food. Several invertebrates such as earthworms, meal worms, and crickets are commercially available and are considered a standard food item by many aviculturists. Again, this employs the motion technique and has proven invaluable on many occasions. A few worms, 8 or 10, can be sprinkled onto the feed to attract the bird. If the bird still does not pay attention, try dropping single worms in front of him or tempt him by presenting a struggling worm between your fingers or with a pair of blunt-tipped forceps.

Still another technique that makes direct use of motion requires that enough water be added to the food to allow it to be molded into small "bite-sized" balls. Then standing in a darkened area and rolling or dropping these balls of food near the chick will induce him to chase and catch them.

Ducks, being the water lovers that they are, often can be induced to feed by introducing a very shallow pan of water to the brooder and adding aquatic invertebrates or pellets of food. This can be carried a step further by adding enough dry food to make a soupy mixture. If this is done, it should be stressed that unless the birds also have access to clean swimming water, they should not be allowed to get into this soupy mixture as they will get wet and dirty and can then lose their waterproofing and their ability to retain body heat. Losses can then result. Still another variation of this technique would be the addition of *Lemna* to the water. *Lemna* is the generic name given to a group of tiny, free-floating plants common in still water during the warmer months. You may know this group of plants as duckweed or duck meal. It forms a green mat on the water surface and the tiny roots provide a haven for large numbers of small aquatic invertebrates. A handful of this sprinkled onto the water in the brooder, or even to older birds, will elicit almost instantaneous feeding activity.

We found that certain birds, such as harlequins, seemed to be searching above their heads for food. By plastering small bits of moistened food to the walls of the brooder, these birds could usually be induced to feed.

One of the most successful techniques that I have used is to put a 3- to 5-day-old mallard, or similar duckling that is feeding well, with the difficult species. The feeding activities of this bird are usually sufficient to start all birds to feeding.

Contrasting colors often arouse the interest of these ducklings and once they begin pecking at the color they also get the desired food and begin feeding. Brightly colored crumbles of turkey and game-bird starter feeds are available and can be sprinkled on the top of the regular feed. Initially, the floor of the brooder can be covered with a clean white cloth, paper towels, or similar non-skid material, and the regular feed, which is usually of a contrasting color, can be sprinkled over a rather large area of the floor. Visibility of the food is then assured and often the birds will feed readily. Sometimes it may be necessary to employ one of the motion techniques too.

If none of these techniques work, it may be necessary to force-feed the birds to ensure that they get adequate nourishment. This should be done with extreme caution and gentleness. These tiny birds are very fragile and are easily injured. The desired food should be mixed with enough water so that soft pellets can be formed. The bird should then be gently but firmly held and the bill opened and the pellet inserted well into the bird's mouth. Allow the bird to swallow each pellet before proceeding with the next. Usually 5 to 10 of these small pellets, depending upon the size of the pellet or the bird, should be sufficient for any single feeding. Birds usually require 10 to 20 percent of their body weight in food every 24 hours. It is always better to give small amounts several times daily than trying to force the entire amount into the bird at any one time.

Now that we have discussed several methods of feeding these birds, you have undoubtedly asked *what* you should feed them. For most species, one of the commercially available turkey or game-bird starter crumbles with a protein content of 24 to 26 percent is usually sufficient. All sorts of food supplements can be added but care should be exercised to ensure that nutritional imbalances don't result.

We found a mixture of boiled and drained hamburger and egg yolks ground together and sprinkled onto the floor of the brooder was eagerly consumed by goldeneyes, scoters, mergansers, harlequins, and long-tailed ducks. Once they were feeding on this mixture, we began adding increasing amounts of dry crumbles so that they were weaned away from the meat and eggs within 7 to 10 days.

Trout chows have been developed for use in fish hatcheries. Fish meal is the usual protein base in this feed and it has been successfully used to rear sea ducks.

Live foods of a large variety are available but should be used sparingly. Don't make these items the major portion of the diet but use them to get the birds eating; then gradually wean the birds away from them.

As with the techniques for getting birds to begin feeding, the items fed are numerous and everyone has his own favorite formula. The amateur will find that he has better luck if he uses the regular formulas and leaves the specialities as a challenge to his increasing knowledge in future years.

PART VII

General Care of Ducks

MELVIN BLOCK

Artificial Lighting

Black Duck

The breeding cycle of ducks is tuned to light. Length of days plays an important part of the breeding process. Experiments should be done with care since ducks can easily be thrown into molt.

There are two systems of lighting used by good management under various conditions.

#1—Morning lights to make a 15-hour day and
#2—Evening lights to make a 15-hour day.

We fell that by using the dual system of part morning and evening lights it will co-ordinate a near-natural system of balancing the natural daylight period when early May arrives. (See chart.)

When the lights come on in the morning, food and water (if no pool is provided) should be available. If this procedure is neglected, most of the artificial light benefits are lost.

Depending upon the judgment of the management, artificial lights should start in early January, increasing the light length by degrees (see chart) up to the latter part of February and maintained to early May. This principle in lighting can be used to a tremendous advantage.

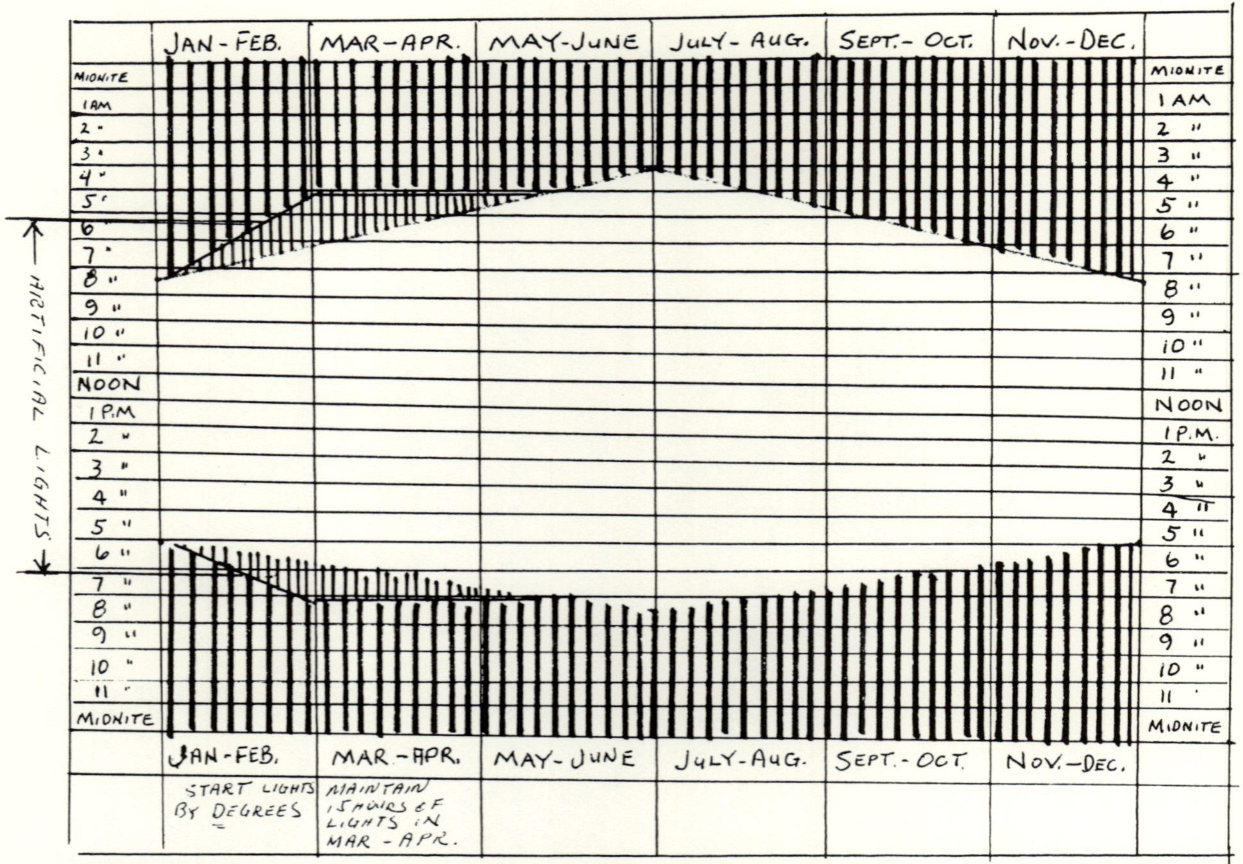

Artificial Lighting

MELVIN BLOCK

To maintain good, vigorous, and healthy birds, the morning feeding should be given very early in the morning. The daily ration is divided into two feedings, the last feeding about four hours before darkness sets in. Birds are creatures of habit and if irregularly fed, they are restless and discontented. If overfed they become fat and lazy; if their food is insufficient, they will not prosper in health and will make poor specimens.

The digestive apparatus of a bird is easily thrown out of adjustment. Supplied with good food it acts with precision, but if neglected it stops and fails to accomplish its regular duties. Regularity in time of feeding and in the quantity of food given is of great importance. The knowledge of the time and manner of feeding and knowing when a bird is fully satisfied must be acquired.

CARL STRUTZ

Winter Care of Wild Ducks in Captivity

The wintering of most North American species of wild ducks is really not too difficult. While they do demand reasonable care and quarters, almost anyone can cheaply provide for their needs during cold weather.

The hardier species such as mallard, black duck, canvasback, redhead, pintail, scaup, ringneck, and wood duck can withstand cold to the zero mark providing they have open water in which to swim and warm their feet. They should be provided with some sort of shelter under which they can retreat during severe cold or snow storms.

This type of shelter can be constructed of 10 mil clear polyethylene over a frame of 2 x 4's. The construction should be made with a solid board for north, west, and east walls, but the south side could be polyethylene covered halfway to the ground. This will allow more sunlight to enter the building and give your birds real protection from the cold. Figure 1-A. The size of your structure will depend upon the number of birds to be kept. Each bird should be allowed approximately one square yard of space.

The accompanying photograph (Figure 1-B) is a good example of winter protection for ducks outdoors with open water.

To protect the polyethylene covering it is a good idea to cover this material over the entire structure with regular window screen which will prevent hail or broken branches from puncturing your polyethylene during summer storms.

In below-zero weather and violent snow storms the birds should be shut in such a shelter and then watered once a day in appropriate watering pans.

The less hardy species, which include teal, shovelers, gadwall, baldpate, mandarins, ruddys, etc., must be kept indoors where the temperature does not get below the freezing mark.

An indoor pen 10 x 14 feet will accommodate about 30 birds without too much crowding. Do not construct the four sides of the pen with solid plywood as this will prevent natural light from windows

FIGURE I—A

A Wintering House

FIGURE I—B

Winter Shelter

brightening your pen. One by 2-inch welded wire panels will provide a fine enclosure. (See Figure 2.)

The building should be heated automatically by either gas, oil, or electricity, whichever is the most economical in the area. The thermostat regulator should be set to control the heat at about the 40- to 45-degree range.

Pens should be bedded with clean, dry straw. Mouldy straw must be avoided as you could destroy your entire flock with mouldy litter.

It is not necessary to clean your pen unless the litter builds up too fast. Once a week, or as often as the top straw becomes messy, add new straw lightly over the floor of the pen. Should there be an excess amount of dampness in your building, necessitating more frequent bedding, ammonia fumes could develop and then of course it is wise to clean out all of the litter and start over again. In some dry areas pens do not have to be cleaned all winter.

Provide two infrared lamps, evenly spaced over each pen. *These lamps should not be closer than five feet above your birds. DO NOT BURN THEM CONSTANTLY* or it will throw the birds into an early molt and ruin all production potentials in spring. However, the lights are needed to prevent feather mould and so it is a good practice to burn them for a period of *THREE HOURS EACH DAY*. A good schedule is to have them burning from 1 P.M. until 4 P.M. daily. This will give the birds adequate time to preen and dry their feathers.

The birds kept outdoors should have open water in all weather. This can be obtained by either constant running water or a good insulated automatic tank heater to keep the surface free of ice. Figure 2-A shows how constant pumping of water will keep a small area free of ice in cold weather.

For birds housed in a heated building the series of photos Figures 3, 4, 5, 6, and 7 show a simple and economical method of providing adequate water facilities for indoor penned birds. Waterfowl like to swim and bathe, so in order to keep them contented and alive, water for swimming is imperative.

A water tank can be made according to specifications by any good tinsmith out of galvanized steel. The one shown in Figure 3 is 24″ x 40″ x 12″ deep and this will accommodate approximately 25 to 30 birds.

To assemble, place a strip of 18″ aluminum tin around the inside of the tank (Figure 3). This can be fastened to the inside of the tank by using regular wood molding and metal screws. This tin barrier is needed to keep the ducks from splashing water all over the pen.

FIGURE 2
Bath Time

FIGURE 2—A
Winter Care

PHOTOS: CARL STRUTZ

FIGURE 3
Duck Bath

FIGURE 4

A Ramp Out of the Tub

PHOTOS: CARL STRU

FIGURE 5

Dripping Area

FIGURE 6

Ramp to Bathing Area

Next, place a small tank ramp made of ½″ hardware cloth inside the water tank. This makes it possible for the birds to get out of the water (Figure 4), onto the drain board (Figure 5), from thence to the approach ramp (Figure 6), and into the pen. The tank ramp (Figure 4) should have legs on one end so as to keep it flush with the edge of the tank. The approach ramp shown in Figure 4 is 14″ x 22″.

Next, build your drain board (Figure 5). It is best to build this with solid sides so as to prevent emerging birds from shaking water all over the pen when leaving the tank. This drain board should be built to slant toward the tank so the dripping water will run back into the pan. The birds will have no trouble at all walking on the plain board since the slant should not be that great.

Now add the longer "approach drain" ramp which will provide the birds with a means of getting down to the strawed pen. (See Figure 6.)

Underneath the approach ramp place tin catching pans which can be made to the exact size of the approach ramp so as the entire space underneath the approach ramp is protected and will catch waste. One single pan can be made if desired. (See Figure 6.) By placing these pans under the approach ramp the pens will be kept dry.

Be sure to close the sides below the approach ramp and the drain board ramp so that the birds cannot get under the ramps and dabble in the filth of the droppings and water in the pans. The birds should be kept on clean straw at all times.

To give the birds side access to the approach ramp it is wise to set concrete blocks, tile or bricks along the one side so the birds will not have to go around to the end of the approach ramp before being able to get up. Then, too, be sure all ramps are against the one side of the pen as this will not take up so much room and the birds can all get to the ramps on one side and on the end. (See Figure 7.)

Before building the ramps, etc., study the photos carefully. It is a very simple construction and can be built by the novice.

Supply feed by a hanging self-feeder placed low enough for all birds to reach and in order to prevent small teal from crawling underneath the feeder. Feed a good clean scratch feed which makes a wonderful winter diet. Occasionally feed greens to your birds in the form of lettuce leaves. It is a good tonic in winter. It is all right, too, to mix a small portion of starter crumbles with the scratch to increase the protein content of their diet.

Fine sand is another important need of birds in confinement and a container with sand should be kept available at all times.

FIGURE 7 *Mandarin Drake Bathing*

PHOTOS: CARL STRUTZ

FIGURE 8 *Pump to Water Supply*

The tank should be filled with water to about 3 inches from the top. Change water at least *TWICE A WEEK* and oftener if needed depending on the number of birds using the tank. Filthy water is a health factor so use good judgment and change whenever needed.

To make the filling and cleaning of the tank a simple matter a small paddle pump can be used. It will draw all the water out of the tank and then when hoses are reversed the tank can be filled with clean water. Silt and all is drawn out by a good paddle pump and these are very efficient and work splendidly for this purpose. Be sure the motor is large enough, however, and it is recommended that a half-horse motor be used to operate the pump since the pump does require sufficient power when water is drawn out and forced out of the building. Figure 8 shows how the motor and pump can be built on a small moveable table.

Specifications of Equipment for 25 or 30 Birds

1. A good steel tank 40″ x 24″ x 12″.
2. A strip of aluminum tin approximately 110″ long and 18″ wide.
3. Tank ramp—22″ x 14″ frame covered with ½″ hardware cloth.
4. Drainboard—24″ x 20″ with solid sides.
5. Drain approach ramp—frame 24″ x 36″ covered with ½″ hardware cloth.
6. Drip pan or pans large enough to protect entire space *below* approach drain ramp.

JOHN ELLIS

Winter Care of Waterfowl

Most species of swan, ducks, and geese can be kept in good health with little or no protection from the elements, even in the northern sections of North America. A certain amount of open water is a must for most ducks and some of the smaller species of geese.

If at all possible, it is much better for all waterfowl to have enough open water for swimming, bathing, and drinking. Birds that have this available to them come through the winter in much better flesh and feather. These birds usually produce more and a higher percentage of fertile eggs than birds that have only a small amount of water and are forced to eat snow during bad weather.

In deep ponds or lakes it is quite easy to keep a large area free of ice even in below-zero weather. All that is needed is a small air compressor which is hooked up to a metal or plastic pipe with 1/16-inch holes drilled into every 8 or 9 inches. Six feet of this pipe laid out straight or formed into a circle will keep a large area open, providing the pond is a fairly large body of water at least 6 to 8 feet deep. The perforated pipe should be weighted down in some way to keep it in place at the bottom of the pond.

If the pond is a shallow one with a fair amount of brook or spring water running into it, sump pumps work best. The submergible type is best and should be placed into a cylinder made of ½-inch galvanized wire to keep leaves and other objects from clogging the pump. From the outlet at the base install a vertical pipe with an elbow so that the outward flow splashes across the top of the water. Each pump will keep about 36 square feet of open water during weather of 10 to 30 degrees, less than that in colder weather.

During the winter, lettuce, endive, cabbage, and other such greens may be fed but are not necessary if the birds have good quality pellets available to them at all times. I feed a poultry developer pellet and no greens during the winter months, supplemented with whole corn. This food is kept available to them at all times. I have tried many types and assortments during the past and find these two feeds about as good as any and this diet will keep all waterfowl, including diving ducks, in excellent condition.

A few species of waterfowl must be kept in winter shelters where the temperature can be maintained at 30 to 50 degrees during the winter months in the northern sections of the country. These species will do only fairly well. Losses and poor breeding results should be expected. The birds that do need these accommodations are the tree ducks, some of the smaller tropical and semi-tropical species of ducks, Oronoco and Ne Ne geese and the radjah shelducks. All of these species have very sensitive feet and legs and are very susceptible to frost. Some strains of coscoroba and black-necked swans cannot take severe cold weather while others do well in the north and can take sub-zero weather with no ill effect. Northern bred birds are desirable.

If small geese and ducks are kept out in areas where deep and drifting snow is expected, some sort of building or shed alongside the water should be provided. This serves as a good dry place for the food to be kept and also serves as a shelter for the smaller birds to use during snow storms.

One good thing to remember is a bird can only be as strong and healthy at the end of the winter as it was when winter began. During the fall months the waterfowl must have all the food they will eat to enable their bodies to fatten up and produce good healthy feathers and down. Good healthy birds will eat much more food in September and October than at any other time of the year.

GUY A. GREENWELL

Waterfowl Predators vs. "Robin Roost"

"Is it necessary to build a high fence around my pond if I want to keep a few ducks or geese?" That question is asked of the waterfowl breeder as often as any other, if my experience is average.

Actually, anyone who keeps wild waterfowl in captivity has an obligation which goes much further. Fencing to prevent the birds from straying and to protect them from harassment is universally needed. There are other measures which can help, too. The nature and extent of that other protection depends upon the kinds of birds involved as well as the location and surroundings.

The story of our efforts here will illustrate some points. Twelve years ago, when we came here, this location became my fourth for waterfowl keeping. It is in the wooded hills which edge the Ozarks, and we were not certain just what predator problems we would encounter. After we'd built a pond a few yards from the end of the house, we installed a 6-foot fence around 4 acres, including the pond and the house. We intended to live with the birds, and we *have* done that.

We obtained a flock of gray call ducks for testing, and our struggle with the predators began.

The list of enemies which may prey on waterfowl is long. Here, it would begin with great horned owls, followed by stray dogs, black rat snakes, cotton rats, opossums, Norway rats, snapping turtles, red foxes, striped skunks, coachwhip snakes, raccoons, starlings, coyotes, gray foxes, stray house cats, common crows, bullfrogs, and mink, in about that order of importance. Some of those bother only eggs or very young birds. There are no hawks on our list. We have never lost a duck to a hawk, with the exception of a mandarin drake killed one winter by a goshawk. That vagrant was the only goshawk we've ever seen in these parts. Obviously, every other duck breeder would have a slightly different list.

Our list is sufficiently varied, however, that we have been faced with several distinct problems. Our 6-foot fence, with the lower half of 1-

inch hexagonal mesh, has stopped the stray dogs and other ground predators which do not climb. When we cleaned a strip around the outside, laid an 18-inch strip of hex mesh wire flat on the ground, laced it to the bottom of the fence at ground level, and let the grass grow up through it, the fence became effective against skunks and other diggers.

We considered the need for electrifying the fence to discourage the climbing animals. That has been done very successfully by installing charged wires on insulators halfway up the fence and near the top on the outside. However, we were interested in learning what happened on the pond at night, so we tried night lighting.

We installed two 250-watt flood lamps on the end of the house, to shine out over the pond and allow us to see what transpired. From my bed, I could see much of the pond and the fence line along the far side, 80 yards away. Within 4 months we had killed 6 great horned owls and lost 6 gray calls.

We installed two more flood lamps on a pole in the back yard, to light a greater area and increase the level of lighting over the pond. We set up some pole traps to reduce the owl population, or so we thought.

The level of lighting which the four lamps gave was effective in inhibiting the entry of raccoons, foxes, and possibly mink. We began to build a collection of wild ducks and geese. They would be able to escape the owls by diving, and we were anxious to get started.

As it developed, the gray calls proved to be good buffers for the wild waterfowl against owls. Owls were not effectively restrained by our lighting and we lost birds, but most often they were call ducks. Their slower reactions made them more vulnerable than the quicker, and more costly, wild birds.

We added another light on a pole to shine across the valley along the upper fence line. That permitted concentrating the beams of the other lamps over the main pond and fenced area. We began using spot lamps instead of floods for some of the lights. Those changes seemed to inhibit the owls, so that instead of coming into the area in the low, headlong dash which is their characteristic striking maneuver, they paused to perch at the edge of the lighted area. They would stare fixedly at the huddle of ducks in the center of the pond, then perhaps move to another perch. More often than before, they would fall victims to our pole traps or a scopesighted .22.

As time passed, we were losing relatively few ducks, but although our records showed we had eliminated 170 owls, we had not greatly altered the frequency with which they appeared. When we eliminated 2, it was only 4 to 8 weeks before we would hear or see 2 more. Sometimes we would hear them calling in the woods above the house for a week or two before one ventured into the light. We had gradually changed the flood lamps to spot lamps, directing their beams toward the far corners of the fenced area, well above the water. Now we were losing very few ducks, but our birds were enduring fairly frequent frights. By glancing out the window, we could tell at once if they'd seen an owl, by noting whether they were bunched near the center of the pond.

One night, I shot at an owl perched in the big elm across the pond, cut some feathers, but missed the body. The frantic flight of that bird out of the light gave me an idea which led to an interesting interlude. By that time, we suspected that the appearance of new owls in pairs was a result of territoriality—that each pair defended its hunting territory against other owls, and it was only when we removed one bird or a pair that a new bird or pair moved in. We began a campaign, rousing from bed every time the ducks signaled danger and attempting to shoot very close to, but not kill, either of the owls which lived with us at the time.

By the time, too, the fact that the pole traps took an occasional kingfisher, robin, sparrow hawk, or mockingbird had created some pretty sticky situations between my wife and me. We had quit using the traps.

The fright campaign did not progress rapidly, for owls do not learn quickly. It was months before we felt we'd accomplished our purpose. We'd made our territorial pair of owls afraid to come into the lighted area. *Then we enjoyed two fine years without losing a duck to an owl.* Sometimes we'd hear them calling in the trees not more than a hundred yards behind the house. Occasionally we would see one at dusk, perched on a dead stub up the valley, but the taboo held.

Then one night I was awakened by the thunderclap splash made by more than a hundred ducks diving simultaneously. A newcomer horned owl, small and dark in appearance, swung back and forth over them, putting them down repeatedly. Apparently some benighted hunter had killed one or both of our house-broken owls. We had to begin again.

Our landlocked ponds are not on a stream channel. Our choice of this location for a waterfowl area was deliberate. Along the streams in our area, raccoons and minks can present real problems. Too, the spring-fed stream valleys, draining our eroded hills, are subject to flash floods. Floods can load one's fences with debris, sweep away sections of them and release all of one's birds. Such deadly predators as snapping turtles and mink present a greater problem along streams, too. With flowing water, special structures are needed to keep them out.

Very early we decided to keep a mowed strip 12 or 14 feet wide outside the fence. It has helped, not only in exposing the activities of animals outside the fence but also in reducing the time which they spend along the fence. Our lighting has let us see that for ourselves.

Inside the fence, we leave strips and clumps of grass to mature to provide nesting cover for those ground nesting ducks which are reluctant to lay in boxes for us. Between the high grass patches, we have strips of short grass, radiating in direction from that bedroom window.

We have been able to watch female ducks sneaking toward their nests and thus find them more quickly, but mainly the high percentage of open ground has proved effective in inhibiting residence by cotton rats and snakes. Also, the mowing keeps tender young grass coming along for the geese.

Our system is far from perfect, but it has the advantage of being less costly than most. We have never had to kill a fox or coyote, though we've seen coyotes and both species of foxes outside the fence. We haven't seen a raccoon within the fence for 8 years. We have hauled away a few black rat snakes and opossums and released them miles away. We have to conduct a Norway rat program for a couple of weeks about twice a year, cleaning up burrowing sites and fumigating burrows. That is all the killing we have to do. We live-trap a few stray cats and take them to the Humane Society.

Our wood and mandarin duck nest boxes are elevated a foot or two above the ground, with rough board ramps up to the openings. We have had starlings build nests over clutches of duck eggs and have had them puncture two or three. We once found a stray cat with a new litter of kittens atop another clutch. Recent research indicates that if one dyes the sawdust with which he floors his nest boxes bright red, starlings will have nothing to do with them. We will try that soon.

There have been some bonuses which we have come to understand were a result of our lighting. Wild wood ducks began to come to feed with our birds at dusk, remained to roost with them, and left again at dawn. They came from a small river named, inappropriately, "Shoal Creek," which passes a mile away. During the first summer we were here, we could find only three broods of young woodies along 10 miles of the river. Each year more woodies came to us. Fall populations have peaked at about 200 birds coming in to roost and feed. Spring peaks average somewhat lower.

The number remaining to nest along the river has increased with several pairs headquartering on our pond during the laying period. The past two summers, 20 and 17 pairs have centered their activities here while the hens were laying. Four hens have walked broods in here to rear them.

We know that wood duck nesting along the river has increased at least 600 percent, without anyone erecting a nesting box. We can only assume that the added protection afforded by roosting under our lights has resulted in much greater survival of breeding stock.

We've had the opportunity to observe nocturnal activities of waterfowl. We've learned that our incubating geese sometimes leave their nests at night to feed and bathe. In fact, some individuals do it habitually. Our ducks and geese interrupt their night's sleep for feeding periods, usually twice during each night. It would be interesting to know if those things happen when there is no night lighting.

Our lighting lets us see what goes on out there and has added to our pleasure from the birds. We have no indication that either molting times or the onset of breeding has been affected. Now, we have a second pair of owls "trained" and the ducks can attend to their normal business.

PART VIII

Special Problems in Raising Ducks

J. D. WALLACH

Diseases of Waterfowl

Blue-Winged Teal

Raising and maintaining waterfowl either for profit or personal enjoyment requires a knowledge of the birds, their habits, shelter and nutritional requirements, and a practical acquaintance with the common disease problems that may be found in a flock.

Records of individual birds in the flock are an aid to a breeder and an invaluable tool to a diagnostician. The problem of diagnosing a disease problem is like detective work in that a dead body alone is often insufficient evidence to make a specific diagnosis. Large amounts of background information and detailed laboratory work may be required to "nail down the lid" on a diagnosis.

Daily inspections of the flock and early examination of dead birds is a good way of preventing disease problems from getting out of control. Even if a bird has been killed by a vehicle, a predator, or its mate, a detailed examination is of value. One can determine the condition of the bird to see if the proper state of nutrition exists, find foreign bodies in the digestive tract, or find chronic diseases such as aspergillosis or air sacculitis.

Birds should be examined as soon as possible after death since decomposition masks many of the "clues" left by the disease. It is a valuable asset to establish a working relationship with a state or university diagnostic laboratory so that competent diagnostic service is available.

It is hoped that the following paragraphs will be of some value to waterfowl breeders in describing some of the more common diseases associated with the keeping of waterfowl as well as describing some of the more recent ones.

Intoxications

Poisoning of waterfowl is usually the result of careless management, careless visitors, or sadists. Rock salt left from winter use or the resulting brine allowed to puddle up within reach of the birds is a common cause of salt poisoning. The birds show convulsive seizures and prostration. Salt should normally be in the diet at the rate of 0.5 percent; levels of 2.0 percent in the diets of breeding birds causes a low fertility rate and low hatchability of eggs. This level in the diet of ducklings causes depressed growth.

Other common sources of poisoning are rodenticides and pesticides in a grain carrier, granular fertilizers, and herbicides. Arsenic poisoning can occur with wind-blown orchard spray.

Lead poisoning occurs when birds have access to old oil filters, lead base paint, or lead shot (very common in shooting preserves).

Blue-green algae and toxic terrestrial plants such as bullnettles, cocklebur, or nightshade can cause sudden deaths.

It is essential to have a detailed history when a poisoning is suspected as there are several hundred thousand agricultural chemicals commonly used today.

Frostbite

Under normal weather conditions waterfowl can withstand severely cold temperatures. This is assuming that plenty of dry bedding or open water is available.

The webs of the feet are susceptible to cold injury. The first signs appear as pale areas, then advance to angry red depressed spots surrounded by a definite white border. (Figure 1). Once this has occurred, a partial (Figure 2) or complete loss of the webbing can be anticipated.

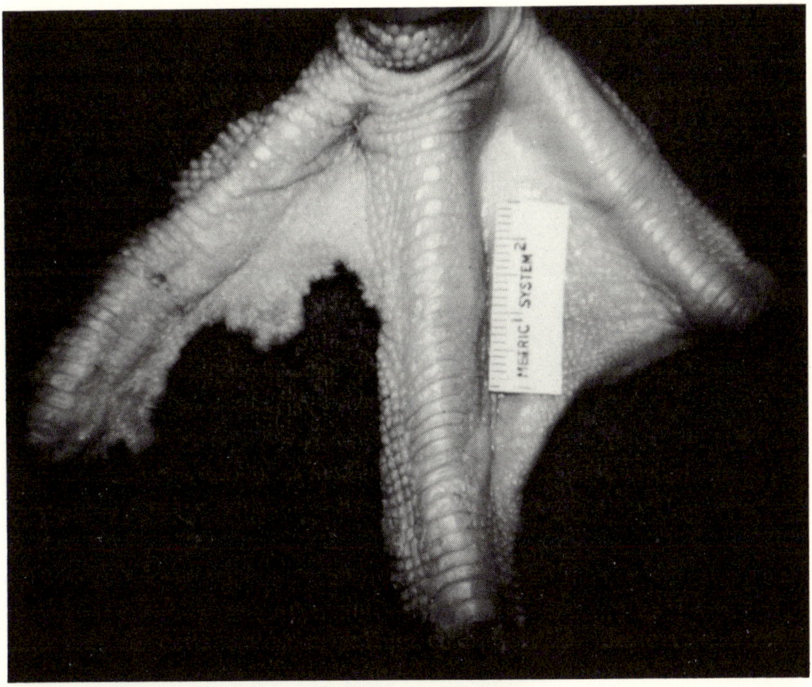

FIGURE 1 *Frost Damage*

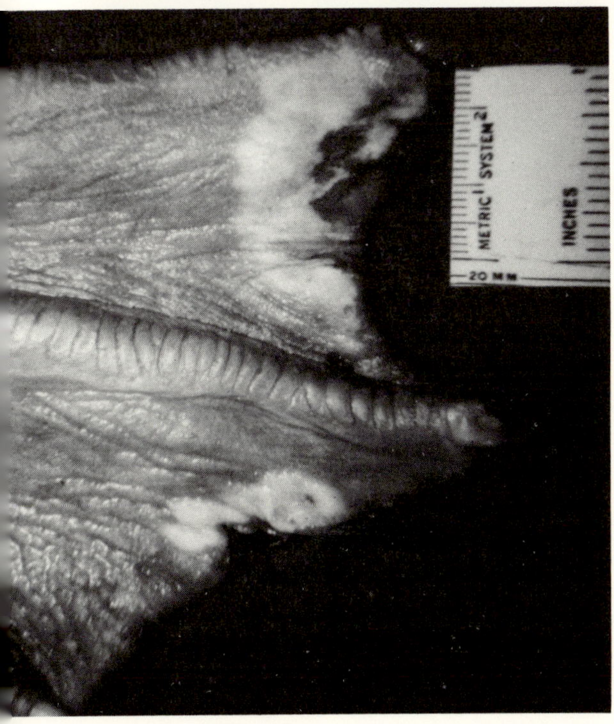

FIGURE 2 *Frost Damage*

PHOTOS: J. D. WALLACH

Sudden losses of otherwise healthy birds may follow the web injury as the injured blood vessels leak bacteria into the general circulation.

Duck Virus Enteritis

Duck virus enteritis, or Dutch duck plague, is a contagious and often lethal disease of waterfowl, causing internal bleeding and severe diarrhea in both wild and domestic ducks, geese, and swans. To date most outbreaks have been confined to local epidemics in the United States in New York State, Maryland, and Pennsylvania, although it has been reported in Holland, Belgium, India, and China.

Waterfowl may pick up the disease by drinking or swimming in water polluted by feces from infected birds or by eating contaminated food. Stresses and poor sanitation may increase the severity of the outbreak. Infected birds become listless, lose their appetites, and develop a great thirst and watery diarrhea. Other signs might be swollen eyelids and drooping wings. The virus is readily killed by heating and disinfectants, and a successful vaccine may soon be developed.

Hardware Disease

This malady, also known as traumatic ventriculitis, may occur with some frequency when curious birds have access to trash dumps, public abuse, or are fed floor sweepings. Small nails, bits of wire, broken glass, golf tees, and fence staples are common offenders. Unless a place is unusually filthy the incidence of this disease is low.

Affected birds exhibit periodic loss of appetite, chronic wasting, and swollen abdomens. The disease may be diagnosed in the live bird by taking an X ray of the abdomen.

On autopsy one may find a generalized infection in the body cavity where the foreign object has punctured the gizzard wall or a local abscess in the gizzard.

Parasites

Although worms are periodically identified from waterfowl they are not often of significance and are therefore omitted from this discussion.

One of the most commonly seen parasites of waterfowl is *Sarcocystis rileyi*. At one time it was thought to be a protozoan disease of

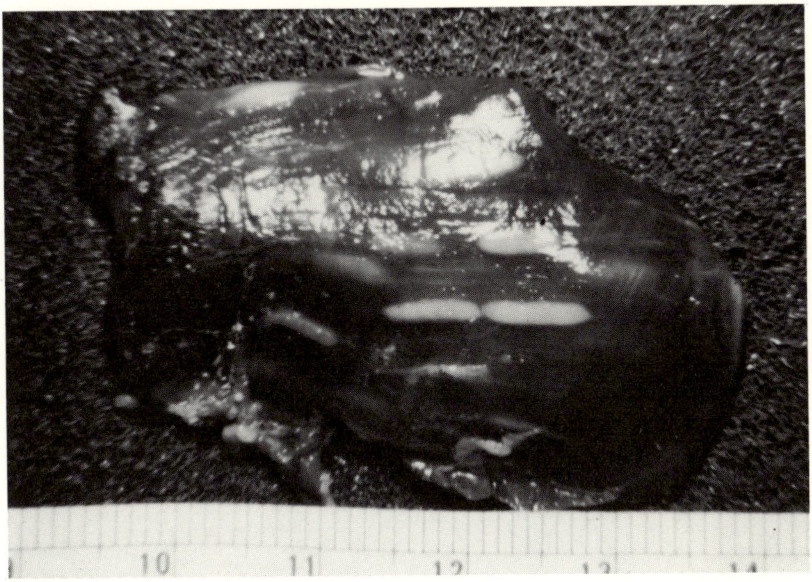

FIGURE 3 J. D. WALLACH

mammals, birds, and reptiles; recently, however, these "parasites" have been identified as a fungus, *Aspergillus spp.*

These "parasites" reside within the striated muscle cells of the heart and breast muscles or appear as white streaks or wheat grain sized nodules. (Figure 3.)

The infective stage is shed through the feces and has been found in food fish. Most birds and animals are not bothered by even heavy infestations. However, a toxic waste product may be produced under certain conditions that can cause emaciation and muscular weakness. There is no known effective treatment and the best preventative is good lake sanitation.

Leucocytozoon is a malaria-like blood parasite of many wild and domestic birds and primarily affects young waterfowl under 12 weeks of age. The disease may be evident only a day or two prior to death. The ducklings stand about in a depressed state reluctant to move. They may gasp in an attempt to get more oxygen. Older birds are often carriers without appearing to be ill; with very heavy infestations they become emaciated and anemic.

The disease usually occurs in mid-summer and is spread by black flies. The entire cycle through the duck and fly is 15 days. Diagnosis is made by finding the parasite in the red blood cells of sick birds. There is no known effective treatment of the disease.

Unhatched Eggs

The examination of unhatched eggs is a valuable asset in keeping track of the health of a waterfowl flock. Poor hatchability may reflect a vitamin B^{12} or protein deficiency in adult birds. "Parrot beaked" embryos indicate a manganese deficiency in the hen. Certain birth defects may reflect the use of improper egg washing compounds.

Bacteria transmitted via the egg, *Salmonella spp.,* for example, may be isolated.

Leg Weakness

Slipped tendon (Syn: perosis) is a nutritional disease resulting from a manganese deficiency. Excessive calcium supplementation to the diet tends to bind manganese making it unavailable.

An outwardly rotated leg associated with an enlarged hock joint is characteristic of the disease. If the disease appears before two weeks of age it is likely that the hen's diet is deficient in manganese. Ducklings and goslings fed a manganese deficient diet will develop the disease in 2 to 10 weeks. The affected birds walk on their hocks or thrust one leg forward and the other to the rear.

There is little that can be done for ducklings with advanced perosis. It may be arrested in the early stages by supplementing the diet with manganese. If there is any incidence of this at all in the flock, a hint should be taken to add more manganese to the diet.

Calcium deficiency (Syn: nutritional secondary hyperparathyroidism, fibrous osteodystrophy, osteomalacia) is a problem which occurs in ducklings when their diets are poor in calcium or too rich in phosphorus. The birds may appear still and reluctant to move. A breast abscess or blister often develops because the birds rest on their keel. Sudden deaths in older birds and thin shelled eggs are signs of calcium deficiencies.

Soft bones and enlarged parathyroid glands are post mortem findings. Most commercial feeds have adequate amounts of calcium and the correct balance between calcium and phosphorus. Problems most often develop in flocks in which the feed is a home mix.

Rickets (vitamin D deficiency) results from a lack of vitamin D which is required for the normal absorption of calcium from the gut and deposition into the bone. The addition of sulfur into the diet may indirectly cause rickets by interfering with the absorption of vitamin D.

There are two forms of vitamin D—D^2 found in irradiated yeast is not usually satisfactory for birds; D^3 found in fish liver oil and irradiated animal oils has a high anti-ricket activity for birds.

Ducklings reared in a sunless shelter require more vitamin D^3 than do those with access to sunlight.

First symptoms are lameness, retarded growth, bent or twisted breast bones, and thin shelled eggs.

Supplementation of the diet with vitamin D^3 for two weeks will reverse the symptoms in two to four weeks unless in the advanced state.

Vitamin E Deficiency

Oily supplements added to the diet, such as vitamin A and vitamin D and fish meal or oil often bind with and reduce the activity of vitamin E, producing a deficiency also known as encephalomalacia or white muscle disease.

The symptoms appear suddenly in ducklings most often during the second or third week. The birds may walk with their feet overlapping and show difficulty in rising. Dead birds show pale areas of muscle in the gizzard. During the warm summer months vitamin E activity in the feed falls rapidly; feed should not be ordered or mixed for more than a two week period.

To reverse the high disease levels, vitamin E should be fed for ten days to two weeks, then reduced to normal levels.

Iron Deficiency Anemia

Also known as round heart disease, this is a common problem in birds reared on wire or concrete floors. This is the result of the total diet being deficient in iron and the birds not having contact with the soil. This may especially be true when fresh greens are fed; this food is highly palatable because of its succulence but is deficient in iron. (When too large a percentage of the diets of young waterfowl is fresh greens, caloric exhaustion often occurs. This is because the young birds are not able to efficiently convert raw plant material, green chop, lettuce, and hydroponics food, to energy. It is ironic that the ducklings run out of "fuel" even while stuffing themselves.)

Affected birds may gasp for air, show leg weakness or nervous symptoms, or die suddenly after a stress such as being released into a new lake.

Autopsy will reveal a rounded right ventricle of the heart. This is the result of an overtaxed heart pumping a watery "strawberry pop" blood.

This may be prevented or treated by supplementing the diet or drinking water with iron.

Gout

This is a metabolic disease of a variety of animals including waterfowl. It is usually associated with high protein diets and dehydration. The signs are sudden deaths in the visceral form and chronic lameness in the arthritic form.

In the visceral form peach-colored chalky crystals are found adhering to the surfaces of organs. The crystals are a waste product of protein metabolism.

The arthritic form is characterized by these same uric acid crystals in the joint.

The incidence normally is low but heavy losses may occur on occasion. The incidence is often higher in laying hens. The disease can often be arrested by reducing the protein level of the diets.

Bumble Foot

This is a bacterial tumor-like abscess of the foot. Usually the heel pad is affected but it may also occur on any toe. (Figure 4.)

This is a sporadic local infection, usually of older birds, that may occur on one or both feet. Bumble foot is found most often in heavy birds and is usually associated with some initial injury to the foot pad followed by a secondary infection. *Staphylococcus spp.* are the most common offenders although other organisms have been isolated. In advanced cases the entire leg may show swelling.

Prevention consists of good sanitation, clean litter on rough concrete or gravel walkways, and reducing the feces build-up around feeding and watering areas.

Treatment consists of opening the tumorous enlargement and scrapping out the cheese-like pus, then packing the resulting wound with

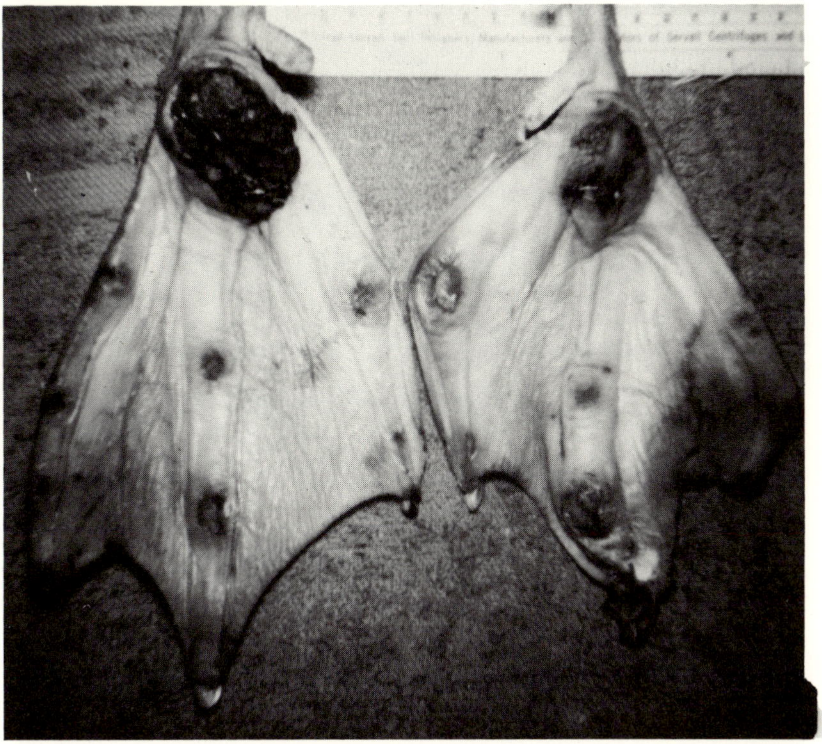

FIGURE 4 J. D. WALLACH

iodine-soaked gauze and giving local treatment with sulfa drug or ammoniated mercurial ointments.

Aspergillosis

This is a common fungus infection of waterfowl also known as brooder pneumonia, mycotic pneumonia, pneumomycosis. The trachea, lungs, and air sacs are the usual site of infection. The *Aspergillous spp.* group of fungi are common air-borne organisms that grow well on stored grain and grain products. The disease is more devastating to ducklings than to older birds. A severe epidemic can develop in brooders that are inadequately ventilated.

Ducklings become debilitated, prostrate, and gasp for air. Very often the birds have a telltale bad breath. The disease is introduced to older birds by moldy feed or litter.

Tumor-like nodules develop in the lungs of young birds; if they survive sufficiently long enough the nodules become laminated with concentric rings similar to an onion. In many cases actual colonies of the organism may be found in the air sacs.

Feeds contaminated with *Aspergillous flavous* may contain a toxic waste product known as aflotoxin. Aflotoxin in relatively large amounts causes sudden deaths in ducklings from liver damage. If minute amounts of this toxin are eaten over a period of several months a form of liver cancer will develop.

Prevention of this disease complex is most advantageous since few treatments alter the course of the disease in advanced forms. Good sanitation and ventilation are essential. Moldy litter can be safely treated with 1 percent copper sulfate solution.

Botulism

This problem also known as limber neck, bulbar paralysis, western duck disease, alkali disease, or lake disease, is a symptom of poor water management conditions. The incidence is usually limited to the hot months of the summer. It is one of the largest causes of death in wild and semi-wild waterfowl.

Decaying organic matter such as water plants, moldy straw, nesting material and bird carcasses make an ideal medium for the proliferation of the causitive bacteria, *Clostridium botulinum.* Although a normal inhabitant of the soil and mammalian intestines, the bacteria does not usually generate its powerful toxin known as botulinum toxin unless conditions become suitable (low oxygen levels in the soil or water because of decaying organic matter).

The toxin is invariably taken orally by waterfowl. This is done by directly consuming infected material or by eating maggots that have developed on the contaminated organic matter. The toxin exerts its effects on the nervous system causing a flaccid paralysis, deranged equilibrium, loss of vision, and death by respiratory paralysis.

The disease leaves no clues in the carcass. Diagnosis can be made from the history and appearance of the dying birds. One can only prove the disease's presence by demonstrating the presence of the toxin.

Prevention is the ideal manner of control. This can be accomplished by a routine daily inspection of the waterfowl lakes and the removal of dead birds and plant material. Aeration of the lakes during the hot summer months is important. Carp and catfish aid in aerating the bottom.

An antitoxin is commercially available for vaccinating and treating waterfowl. A yearly vaccination program may be of benefit in areas of high incidence (prevention is preferred).

New Duck Disease

This is a disease of ducklings characterized by the formation of a false membrane on the surface of the liver and heart. Sometimes referred to as anatipestifer infection, serositis, it is one of the more important diseases of ducklings in the United States. Mortality is usually high. Affected birds cough, have a green diarrhea, and exhibit incoordination. The birds usually show advanced dehydration. The disease is thought to be caused by a bacteria called *Pasteurella anatispestifer;* however, some investigators feel that other bacteria may be involved.

One sees a false membrane associated with the liver and heart; sometimes the trachea, lungs, air sacs, and brain may become involved. The disease symptoms, autopsy clues, and positive bacterial cultures are required to establish this diagnosis.

Sulfa drugs have demonstrated some efficiency in controlling outbreaks. Low levels of antibiotics in the feed apparently do not prevent the disease. Individual injections of an oxytetracycline protects ducklings for a few days during an outbreak.

Salmonellosis

This bacterial disease of a variety of domestic and wild animals is also known as keel or paratyphoid. There are numerous species of this bacteria; however, two in particular are important to waterfowl breeders. *Salmonella typhimurium* and *S. anatum* occur with some frequency in ducklings under crowded conditions. Carrier birds often do not appear to be ill. The active disease usually occurs after some environmental stress; therefore exhaustion and dehydration and inadequate diet during shipment or holding should be avoided.

Ducklings with salmonellosis are dehydrated, pasty around the vent, and may die suddenly. The mortality is high in young birds. In older birds a chronic wasting form occurs with a drop or cessation in egg production. It is important to remember that this particular disease is commonly transferred through the egg. Religious fumigation of the incubator is very important because of egg transmission; the organism usually does not penetrate the egg until the eighth day of incubation. Therefore fumigation with potassium permanganate and formalin on the third day as well as between hatches may be of help in reducing the incidence of this disease.

The clues at autopsy usually include a cheesy cecal plug and small abscesses in the liver. The yolk sac may or may not be absorbed. Isolation of the bacteria from the infected material is required to make the diagnosis.

Treatment with sulfa drugs and oxytetracycline does not eliminate the Salmonella organism but may arrest the course of the disease; in some cases furazolidone (NF-180) has shown a better batting average.

Fowl Cholera

This is a bacterial disease caused by *Pasteurella multocida.* Losses are usually sudden and most severe in ducklings over four weeks of age. The skin may appear red and feel hot to the touch. Death usually occurs at night after feeding contaminated material. Dense populations tend to contract the disease most often. The incidence is usually associated with cold, wet, fall and spring weather.

Fowl cholera spreads by direct contact or through contaminated feed, water, or litter. Soil in contaminated pens or rotting carcasses may be infected for as long as three months. Drying, exposure to sunlight, and disinfection with most commercial products kill this bacteria.

At post mortem the livers and spleens show pin point abscesses; the spleen may be five times normal size. A generalized infection of the body cavity is often seen. Pinpoint hemorrhages are often found on organ surfaces.

No drug available effectively eliminates the fowl cholera bacteria. Losses often occur again after the cessation of treatment. Sulfa drugs in the drinking water control the mortality rate.

Commercial vaccines appear to have some beneficial effects during mild outbreaks.

Air Sacculitis

This disease (Syn: CRD) is caused by a virus-like bacteria, pleuropneumonia-like organism (PPLO) called *Mycoplasma gallinarum.* This is a disease common to all birds, domestic and wild. The organism affects the respiratory tract. In young ducklings the course of the disease is progressive resulting in high mortality after a few weeks; respiratory distress and emaciation for several months usually precedes death in the older birds.

The clues at post mortem are quite characteristic. A white to yellow cheesy material is found in the air sacs. A yellow slime may be associated with the more solid material. A bacteria normally found in the digestive tract, *Escherichia coli,* is a common secondary invader to the original PPLO infection.

Treatment is difficult at best except in the very early cases prior to the secondary infection by the bacteria. Tylocine (Tylosin injectable or via the drinking water) appears to be the most efficient product for treating PPLO infections at this time. After treatment, arrested lesions may be found in older birds.

Duck Plague

This is a devastating virus disease of waterfowl that was once limited to the Netherlands. It is also called Dutch duck disease and, incorrectly, fowl plague. The disease was first identified in the United States in 1967 on Long Island, New York.

The disease has a brief course of a few days with a mortality rate of 90 to 100 percent. The disease spreads by contact with infected birds or indirectly with contaminated ponds. Three or four days after infection ducks become listless and sit with drooped wings reluctant to move or swim. The affected birds have an ocular discharge, labored breathing, and a watery diarrhea. Because of the brief illness the birds are invariably in good condition at the time of death.

At post mortem one finds the organ surfaces covered with pinpoint hemorrhages and false membranes associated with the esophagus and trachea. It takes detailed laboratory work to differentiate this from true fowl plague (which has not occurred in the United States since 1929).

Ducks that survive natural or experimental infections with this virus are immune. Control of this disease is limited to vaccination after an outbreak; the mortality rate usually drops rapidly after vaccination.

Virus Hepatitis of Ducklings

The disease was first described in 1950. The virus affects young ducklings from three days to three weeks of age. Mortality runs about 80 percent. Older birds do not show signs of the disease but may be carriers.

The course of the disease is very brief, going through an entire flock in four days. The incubation period is only 48 hours; one may not see the ducklings sick until one hour prior to death.

At post mortem the livers are enlarged and the surfaces covered with pin point hemorrhages.

The injection of serum in very young birds gives sufficient temporary immunity during the critical age periods.

Bibliography

1. Anon., The Merck Veterinary Manual, Merck & Co., Rahway, N.J., Tnd E. (1961).
2. Hagan, W. A. and Bruner, D. W., The Infectious Disease of Domestic Animals, Comstock Publishing Associates, Ithaca, N.Y., 4th Ed. (1961).
3. Jansen, J., Duck Plague, The Journal of the American Veterinary Medical Association, Vol. 152 (April 1968).
4. Leibovitz, I. and Hwang, J., Duck Plague in American Anseriforms, Bulletin of the Wildlife Disease Association, Vol. 4 (Jan., 1968).
5. Locke, L. N., Irby, H. D., Bagley, G. E., Histopathology of Mallards Dosed with Lead and Selected Substitute Shot, Bulletin of the Wildlife Disease Association, Vol. 3 (Oct. 1967).
6. Reilly, J. R. and Boroff, D. A., Botulism in a Tidal Estuary in New Jersey, Vol. 3 (Jan. 1967).

J. D. WALLACH

Recommended Treatments for Diseases of Waterfowl and Game Birds

Type	*Location*	*Treatment*
PARASITES:		
Sarcocystis (*Sarcocystic rileyi*)	Breast muscle	None
Lice	Skin & Feathers	Dust with DDT, rotenone or malathion.
Scaly Leg Mite (*Knemidokoptes mutans*)	Under leg scales	Dip affected leg in 1 part kerosene and 2 parts raw linseed oil.
Coccidia (*Eimeria spp.*)	Intestine & cecum	Sulfamethazine or sulfaquinoxaline in drinking water (for prophylaxis—nitrofurazone of nitrophenide may be fed continuously).
Black Head (*Histamonas meleagridis*)	Cecum & Liver	For prevention and treatment use dimetridazole and furazolidrone.
Cecal Worms (*Heterakis gallinae*)	Cecum	Phenothiazine
Gape Worms (*Syngamus trachea*)	Trachea	Feed thiabendizole or dust with barium antimonyl tartrate.
NUTRITIONAL DISEASES:		
Slipped tendon (Manganese def.)	Hock joint	Supplement manganese to diet.
Osteomalacia (Calcium def.)	Long bones	Supplement calcium to diet.
Rickets (vit. D^3 def.)	Ends of long bones	Supplement diet with vit. D^3.
Encephalomalacia (White Muscle Disease—vit. E def.)	Muscle & gizzard	Supplement diet with vit. E.

Type	*Location*	*Treatment*
Anemia (Iron def.)	Heart & blood	Supplement diet with iron.
Fatty liver (Choline def.)	Liver	Supplement choline to diet.
Curl toe paralysis (Riboflavin def.)	Toes	Supplement B² to diet.
INFECTIOUS DISEASES:		
Bumble foot (*Staphylococcus spp.*)	Heel pad	Local surgery
Aspergillosis (*Aspergillous spp.*)	Air sacs	Wet litter with 1% copper sulfate
Botulism (*Clostridium botulinum*)	Paralysis	Pick up dead birds and plant material; vaccinate with antitoxin.
New duck disease (*Pasteurella anatipestifer*)	False membranes on viscera	Sulfa drugs
Salmonellosis (*Salmonella spp.*)	Cecum & liver	Furazolidone
Fowl cholera (*Pasteurella moltociad*)	Liver & spleen	Sulfa drugs & vaccine
Air sacculitis	Air sacs	Tylosin
Duck plague (Virus)	Pinpoint hemorrhages & false membranes	Vaccinate after outbreak.
Hepatitis of ducklings (Virus)	Liver swollen; pinpoint hemorrhages	Vaccinate ducklings.

JOHN ELLIS

Netting and Safe Handling

Blue-Winged Teal

A suitable net is very important for catching up ducks and geese without injuring them. Nets made for fishermen are very good, providing the net itself is made of fairly heavy cord and not too coarse a mesh. Nylon nets are the least desirable as these can cut and injure large or small birds easily.

I have tried many types of nets. The best type I have used is a net made of heavy cotton cord 18 inches in diameter and 24 inches deep. This net is mounted on a sturdy steel-wire hoop which is attached to a wooden handle 3 to 5 feet in length. There are some nice lightweight aluminum nets also which are ideal for the person who only uses a net occasionally. These lightweight nets do not stand up well with frequent use so are not advisable for someone who uses a net often.

The easiest and safest way to net a duck or goose is to drive the bird along a fence or other barrier and place the net in front of the bird. With time and practice one becomes very good and can usually net the bird on the first try. Netting a bird over the head and body can be dangerous as you may hit him on the head or back with the metal hoop.

Banding Work

The best way to remove a bird from the net is to hold both wings where they join the body with your thumb and forefinger. When holding a bird in this manner do not apply too much pressure as the wing muscles can be injured very easily, especially on small species. Never attempt to carry a bird held in this manner as you may permanently damage the wing muscles.

If the bird that has been netted and removed from the net has to be carried any distance observe the following precautions. (1) Run your other hand along the belly until the legs are between your second and third finger and third and fourth finger or fourth and fifth finger on larger birds. (2) Apply enough pressure to the legs to keep the bird from kicking. (3) With the palm and fingers of your other hand hold the wings tightly to the bird's body. (4) Bring the bird close to your chest if it is small or to your side under your arm if it is a large duck or goose. (5) Always hold the bird so that the head is pointed toward your body or on larger birds, under your arm toward your back. If birds are carried in this manner they struggle much less and cannot strain themselves or get hurt.

In catching swans it is much easier to corner the bird and grab both wings at the shoulders using your two hands. Do not carry a swan any distance by holding him as the wings can be injured very easily because of the size and the body weight of the bird.

Catch and handle your birds as little as possible and only when absolutely necessary. When a bird is netted and caught, regardless of how tame it is, there is a great physical strain and the heartbeat increases greatly. Some birds become so disturbed that you can actually feel and hear their hearts beating.

Driving Ducks into a Trap for Banding

PHOTOS: ED BRY

KERRY A. MULLER

Trapping Escaped Ducks

Everyone who keeps birds will, in all probability, eventually have to face the problem of capturing an escapee. If the birds are pinioned or wing-clipped, this may be a simple procedure; however, if the birds can fly it may become a very serious problem, all too likely resulting in the permanent loss of a highly valued specimen.

When you discover that the bird has escaped, your first question is, "What do I do now?" If the bird can't fly and is nearby, on land, the first thing to do is get your hand net. Everyone who keeps birds should have at least one hand net available in the bird area at all times. The best and most available nets for ducks are fish-dip nets with a 12-inch, or larger, diameter. These can be obtained at any sporting goods store and work very well; however, most have a rather short handle and you will have to make an extension of at least 36 inches.

A second type of net that is excellent for this purpose is made of fine olive-green, nylon mesh and is sold at most pet stores for use in catching small birds. These nets are very useful in capturing the smaller ducks and game birds and have an advantage in that the birds can't become entangled in the mesh.

A very useful and often necessary addition to the hand net is a long detachable pole that can be used as an extension of the handle. This can be either a long, light wooden pole such as bamboo that can be taped with masking tape to the net handle, or even better, an 8 or 10-foot section of aluminum electrical conduit piping, ½ inch in diameter, which is very light and strong.

The second step in recapturing the bird, after getting the net, is to quietly walk behind the bird and attempt to herd it into a fence corner or other confined space. In this operation two people are better than one. Once the bird is in a corner, *carefully* converge on it and slip the net over its head. Once the bird is in the net, try to keep enough pressure on the net against the ground so that it can't force its way under the net rim and get it in your hands and out of the net without delay. When you are removing the bird from the net, take care that his head, neck, wings, or legs are not entangled, or else you are likely to break a bone. When you are wielding a net, remember that the idea is to get it over the bird, not to use it as a club or tennis racket. Birds are often killed or maimed by over-enthusiastic, under-experienced netters. Often, even birds that are able to fly can be recaptured by the above method if care is taken not to frighten them.

If a net is not available, the above method can be used to capture the bird by hand; however, ducks are very quick and adept at turning so they often escape, and additionally there is a strong tendency to grab them by a wing or leg which usually results in a broken bone. If you must capture a duck by hand, first try to secure his neck, quickly following up with the other hand and holding the base of *both* wings. If you miss his neck, try and pin him to the ground firmly but lightly with the flat of one hand until you can secure him by the base of both wings, holding them one between the thumb and first index finger and the other between the first and second fingers. When lifted, the bird's weight should be supported underneath with the other hand. There are other methods of holding ducks, but this is the safest. It is wise to hold the rear of the bird away from your body, as they are apt to be messy. When catching a steamer duck, be careful to not allow the bird to bite, as these birds are quite capable of inflicting painful and sometimes embarrassing injuries.

If the escaped duck succeeds in getting onto a body of water, the problem of capturing him becomes more difficult. If the bird can't fly, he often can be driven to a confined area and captured with a hand net. If the pond can be easily drained it is best to do so; however, if not, and the water is shallow enough, the bird may be driven by men in waders or hip boots. When the bird is in a confined spot a hand net may be placed over him, or for many ducks that can dive the net must be under water and brought upward; for this operation a longer handle is needed on the net. If the pool is narrow enough, often two men on opposite sides can net the duck from shore.

When the bird is on deep water he must be driven by men in a boat or boats. This is a difficult operation at best and with diving ducks can be frustrating. At least two men per boat are needed, one to handle the boat and the other the net. The boat, which should be relatively small and manageable, is best rowed, or in windy conditions a light motor used. The idea is to head the bird into a confined and shallow area so that he is forced to come within the netter's reach. If the bird is not pushed too hard this may be done; however, if he panics he will usually dive, and unless the water is very clear you will be unable to tell in which direction he has gone. In cases of this sort a good deal of patience and perseverance are needed, and the bird may escape in any case.

In some instances a seine net suspended between two boats or worked from two shores may be helpful; however, it is very unwieldy and great care must be taken not to drown the bird. My experience is that it seldom works.

Another method helpful on a smaller pond is that of using a rope, held on each side of the pond by a man. This is initiated at one end of the pond and slapped up and down on the water's surface, the men gradually working their way down the pond. The bird will sometimes leave the water and attempt to escape on land, making him easier to capture.

When your escaped duck can fly, guile must be resorted to instead of force. If its mate is still confined to an aviary, it may be wise to place her in a smaller cage and leave the door to the aviary open. If the bird is used to the area and not too wild it will often return within 48 hours and can be captured upon entering the enclosure by closing the door from a distance with a cord, etc. If the bird remains in the area but will not enter his enclosure, you must then attempt to trap him.

The simplest type of trap, used for a bird that will feed on land, is a wire, open-bottom box. This should be at least 3 feet square and 2 feet high, with wire sides and top. If a solid covering on top is used the bird will probably not enter it.

The trap should be propped up on one side with a stick no more than 8 inches high. This is tied to a cord leading to a concealed location. Grain is then placed near the rear of the box, with a fine line leading out. When the duck enters, pull the cord. It is well to test this arrangement first and if the trap falls slowly add a weight to the front top. A door should be cut in the top of the trap for aid in removal of the bird.

Unfortunately, most ducks are very early and late feeders, as well as being quite wary, so sitting around and waiting for them to go under a trap can be quite tedious. For those aviculturists lacking in time, other alternatives can be considered such as a figure-4 trap.

The figure-4 trigger has the advantage that it can be left unwatched; however, any type of bird or animal can trigger the mechanism and this can be frustrating.

Perhaps the best form of duck trap for the average aviculturist to construct is some form of funnel trap. Of these, there are two basic types, one for dabbling ducks and one for diving ducks.

The construction of a funnel trap for dabbling ducks can be quite varied in size and shape, with one or more funnels; however, the trap should be at least 4 feet wide, 6 feet deep and 2 feet high, and if it is to be used in shallow water where there are tides it should be at least 4 feet high. It must be made of wire, such as 1 x 2-inch or 2 x 4-inch mesh welded wire, or 2-inch or 4-inch mesh chicken wire. If the wire is not self-supporting it must be stretched on a wooden frame. Some species of small ducks can escape from 2 x 4-inch mesh, so this must be taken into consideration.

A funnel must next be made and placed on one bottom side of the trap, approximately in the center. The outside opening should be about 12 inches square, extending along the sides and top inward 12 inches and gradually sloping on top to 6 or 8 inches high at the rear. Neither the trap nor the funnel has a bottom, as in water many birds tend to catch a head or wing in the wire and drown.

To bait the trap grain should be placed in the center and a line extended through the funnel and beyond. Extra grain can be put just to the rear of the funnel for added enticement.

A door should be cut in the wire on top of the trap to aid in removal of the ducks.

The dabbling-duck type of funnel trap can be used on land or in shallow water up to 4 or 5 inches deep; however, if the water is a little deeper make a higher funnel. If you are trapping where tides are prevalent be sure that at least 12 inches of the trap remain above the high-water line, or you will drown the birds.

If the area in which you are trapping has ice or ices over at night, remove the traps until it thaws. Ice will badly crush a trap, especially in a windy area.

At this point a few words of warning should be given: If you are trapping where wild ducks occur, be sure to consult both your state and federal fish and game agents. If you fail to do so you are liable to severe fines and other legal procedures, as you are required to have both federal and state permits to trap wild ducks. In addition, you must be alert for poachers, for if your trap or traps are left unwatched, especially in the early morning and evening, anyone passing by will find the temptation of ready-caught ducks hard to resist. As a final word of warning, anyone trapping birds is liable to find himself the recipient of a delegation from the local humane society, Audubon society, hunting club, or other civic organizations. When that occurs you're on your own.

The second type of funnel trap is that designed for diving ducks. It is in effect like the dabbling-duck type, except that it must be 6 feet high where tides are prevalent as it must be set at a feeding depth of 2 to 3 feet of water at the median water level. One very easy type to make consists of a simple wire cylinder, 6 feet in diameter with a wire top cut 6 feet square and bent over at the corners. The wire can be fastened to itself with hog rings or by bending and a hook left on the top to remove birds. The wire used must be self-supporting and can be either welded wire or 14-gauge galvanized chicken wire, 2-inch mesh. The funnel is like that described for the dabbling-duck trap, except it should be higher, especially when used in mud.

The diving-duck funnel trap can be set in 2 or 3 feet of water from either a boat or by using waders in some areas. Care must be taken to see that the bottom of the trap is secure all around, for if there is a hole in the mud or sand anywhere the birds will find it and escape under the trap. When approaching the trap to remove birds, a baffle must be placed over the funnel entrance. This is simply a pole, jammed in the sand or mud, with a piece of wood or metal attached to cover the funnel entrance. The trap can then be emptied with a long-handled net by opening one side of the top.

Bait that is used can be varied by what is available. For most ducks whole corn is very good, as it is bright and shows up well, does not readily deteriorate in water, and is readily acceptable. If corn is not available, wheat and milo are also good. For some very small or exotic species perhaps other smaller grains will have to be tried.

Some types of ducks don't trap well under any circumstances. These include deep water and sea feeders, such as eiders, scoters, old squaws, and harlequin ducks. Equally difficult are the fish-eating mergansers. If the birds are captive-raised and trained to eat grain, however, they might be trapped. The third type of duck that is very difficult to trap are the stiff-tails, such as ruddy ducks.

There are other methods of trapping wild waterfowl that are used by professionals; however, most of these are not available to the average aviculturist. These include mist netting, rocket or cannon netting, night drives into fixed traps, jack-lighting, and the use of drugs.

Mist nets are fine nylon nets, which are dyed in colors to match their environment and when set properly are almost invisible. Birds cannot see them until they are too close to turn, and they simply fly into them. These nets can't be obtained or possessed without a permit. For most ducks a 4-inch mesh, 60 to 100 feet long, 7 to 10 feet high mist net is used. This is best stretched over a stream or estuary in a shaded place, or used at night. Ducks have excellent vision and any net in the open or improperly set is easily avoided. Often success is obtained by stretching the net under a bridge where the birds habitually fly. Some trappers have achieved results by stretching the net horizontally and baiting beneath it, the net being released manually when the birds are beneath it. Mist nets should *never* be left unwatched and must not be used by amateurs.

Rocket or cannon netting is a technique sometimes used by professionals to capture large numbers of geese, swans, and dabbling ducks on land. This method basically consists of baiting a flat area of ground, and when the birds are in the area shooting a 60-foot-long, 40-foot-wide net over them. The net is pegged to the ground with guy ropes on one side and the other side is connected either to three rockets or three weights which are fired with a black powder charge set off electrically from a distance by the trapper. The net, which is hidden by straw, etc., is pulled by the rockets or weights which are fired simultaneously and the net describes an arc over the flock of birds. This apparatus can't be used without proper permits.

Different types of drugs have been used to dope birds so they can be captured, with varying degrees of success. One drug used is panabarb sodium; however, there is usually at least 6 percent mortality, and this method is not worked out enough to be used with safety. Another drug is plain ethyl alcohol, in which the grain is soaked. This in effect makes the birds drunk; however, there is little control over the amount taken by the birds and if they fly to another area before becoming affected they may drown or injure themselves, as well as being extremely defenseless against predators.

Two methods used to capture ducks by poachers are snares or steel snap traps. Both of these either kill or maim the birds and should never be used.

One basic piece of equipment useful to the aviculturist consists of a wire box, 4 feet wide, 4 feet high and 6 feet long. This should be wire on all sides, including top and bottom, made of a self-sustaining wire, such as welded wire. On one end the entire side is made as a door, which may be lifted over the top. On either side of the door are two pieces of wire 4 feet high and 6 feet long. These can form "wings" which can be folded against the sides or brought out to form any desired angle. This basic trap can be very useful, as it can be set up adjacent to a wall or fence and often one or a group of ducks can be driven into it. This is often much better than using a hand net, and usually prevents badly frightening the birds.

No matter which method you choose to capture your escaped bird, try to remember that his welfare is the primary consideration. If a bird is killed or injured in the process of capture he would have been better off in the wild.

JOHN A. GRISWOLD

Illustrations by Joseph M. Devlin

Methods of Restricting Flight

Redhead Diving

Wherever valuable waterfowl—pheasants, cranes, storks, etc.—are kept in uncovered enclosures, steps must be taken to prevent them from flying. There are five methods that may be employed: feather pulling, wing clipping, brailing, tenotomy, and pinioning. The first three procedures are wholly temporary and the fourth, tenotomy, is not extensively employed. Tenotomy, which involves severing the main wing tendon, the *Tensor patagii longus,* may not be permanent if the operation of burning through the skin and tendon with an electric burning needle is not done precisely. No matter what process is employed for inhibiting flight, only one wing is involved. This is for the purpose of unbalancing the bird.

Pulling a number of flight feathers is the least effective, because these feathers will be renewed within about 8 weeks. Moreover, growing feathers are well supplied with blood, and the follicles from which they grow tear easily. The bird could suffer from a considerable loss of blood if developing feathers are pulled before they are mature.

Wing clipping is useful for temporary restriction of flight, and this method of feather cutting is illustrated by Figure I in which the left wing of a duck is seen from below. A sharp pair of tin shears or a strong pair of scissors is used for cutting the flight or primary feathers.

FIGURE I

Brailing, or the application of a soft leather strap around the flight feathers of one wing, is primarily used by commercial pheasant raisers for keeping young birds in uncovered pens, prior to marketing to the food trade or for release on hunting preserves. Brails must be changed from one wing to the other at relatively frequent intervals. For obvious aesthetic reasons they are seldom, if ever, used by the non-commercial fancier.

The most effective means to render ornamental birds permanently flightless is a simple surgical operation by which the terminal section of one wing is removed. This operation is called pinioning and, in one form or another, is generally in use wherever large collections of waterfowl are kept in confinement in uncovered enclosures.

Because of the many rare birds that must be pinioned at the Philadelphia Zoo we have developed a pinioning technique of our own. Our procedure is slower than more conventional methods, but it has proved most satisfactory. This technique is as follows: A full-winged bird is held with one wing extended and the feathers are carefully plucked from a portion of the wing distal to the thumb (alula). Care is taken not to tear the skin by pulling too many feathers as a group. A few feathers are left on the terminal portion of the wing to provide a convenient handhold while the incision is made. Next, a strong string (we use 36-lb. nylon fishing line) is tied very tightly around the wrist joint behind the thumb, thus stopping circulation from that point to the end of the wing (Figure 2). The plucked section of the wing is then thoroughly wet with a reliable antiseptic (we use a 12 percent solution of Sulfamethazine). Iodine or other alcoholic solutions should never be used. Then a circumferential cut is made beyond the thumb, and the skin and flesh are carefully turned back as far as possible toward the wrist joint, exposing the metacarpal bones (Figures 2 and 3). Next the metacarpal bones are severed with strong shears as close to the base of the thumb as possible,

FIGURE 2

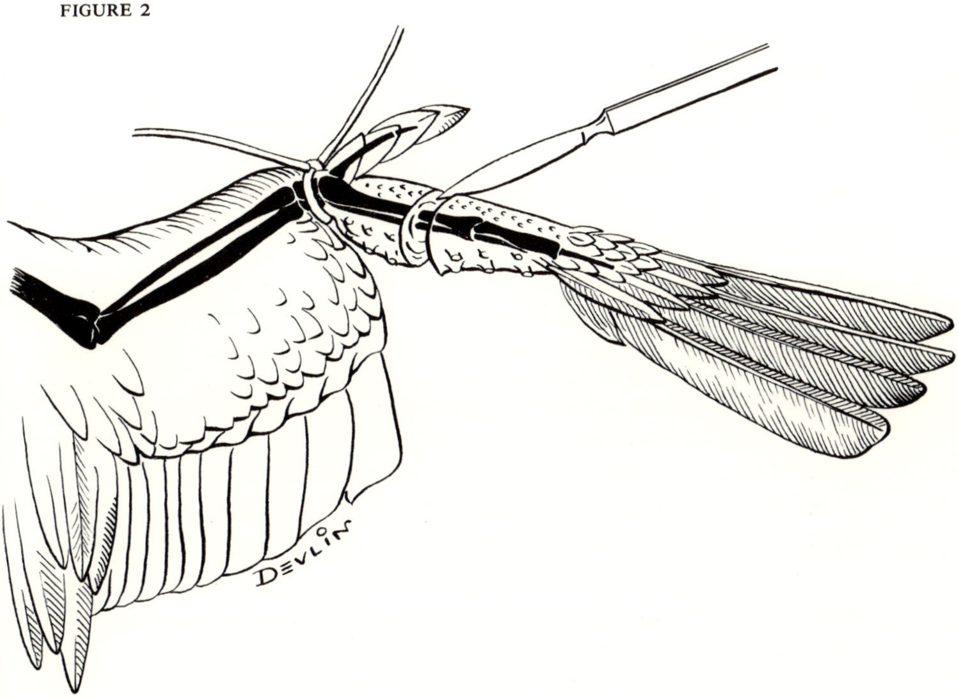

making certain that no jagged ends or slivers of bone remain (Figure 3). The wound is cleaned with the antiseptic solution, and the skin is turned back over the cut end of the bone. Another string is tied tightly about this loose pocket of skin (Figure 4). The excess skin flap beyond this string is cut away, the loose ends of this second string are cut quite short, and the string from around the wrist is removed, thus restoring circulation (Figure 5).

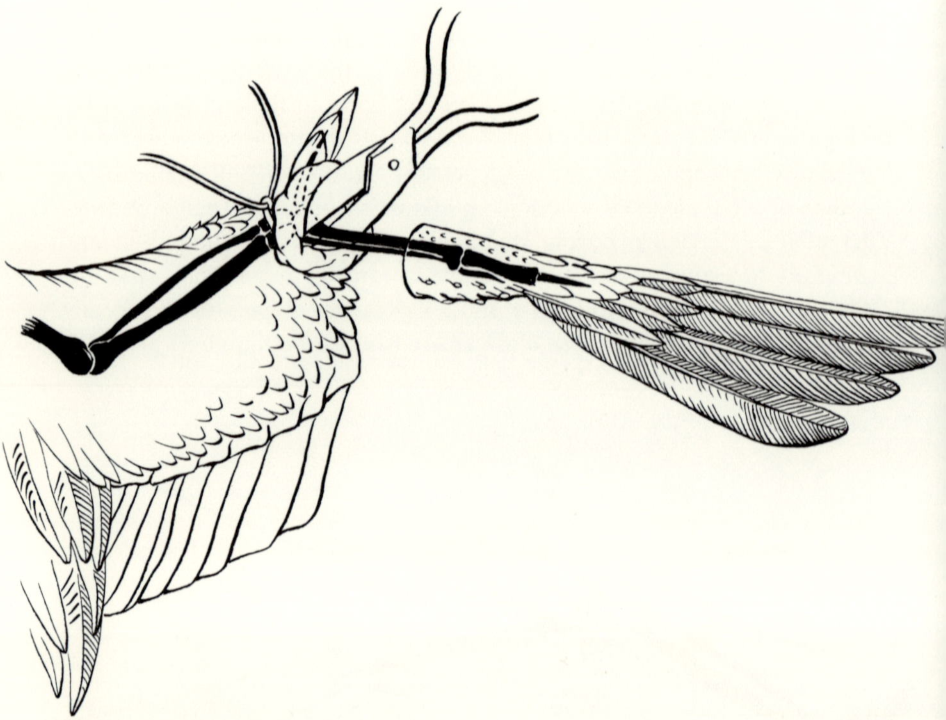

FIGURE 3

Before releasing the bird to its quarters the amputated stump of the wing is watched briefly for signs of bleeding. If the skin has not been perforated or torn during the operation there will be none. The skin pocket over the cut end of the metacarpal bones may swell with blood, but this is of no consequence. The wound heals completely within about two weeks, during which time the string closing the skin pocket falls away as the cut end heals. With a minimum of practice 6 to 10 birds may be pinioned by this method within an hour.

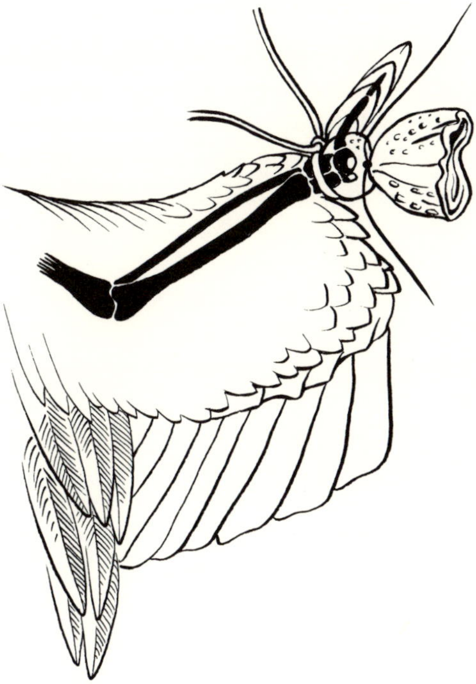

FIGURE 4

FIGURE 5

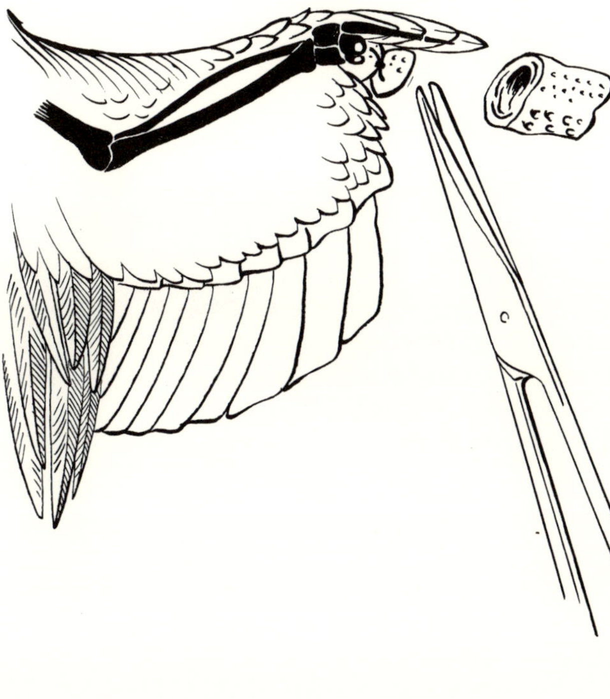

Compared to the usual methods of pinioning, the described process has the advantage of producing a very close pinion, which is especially essential with tree ducks, cranes, and storks. Moreover, there is no chance of the amputated stump being punctured at a later date, because the cut end of the bone is deeply buried in the perfectly healed tissues, which soon are completely covered with well formed feathers. Some pinioning methods require that the pinioned bird be recaught at a later date so that the string may be removed after the blood has coagulated and the wound has partially healed. By using the simple operation just described, a pinioned bird may be returned to its original quarters immediately after the operation, without further worry. Needless to say, a bird in poor condition should not be pinioned, nor should newly acquired specimens be subjected to the shock of pinioning until they are well established in their new surroundings and are eating well.

Adult birds may be pinioned at any time of the year. It has been our policy, however, not to pinion birds during the hot summer months inasmuch as it is unwise to restrain a bird for any length of time when

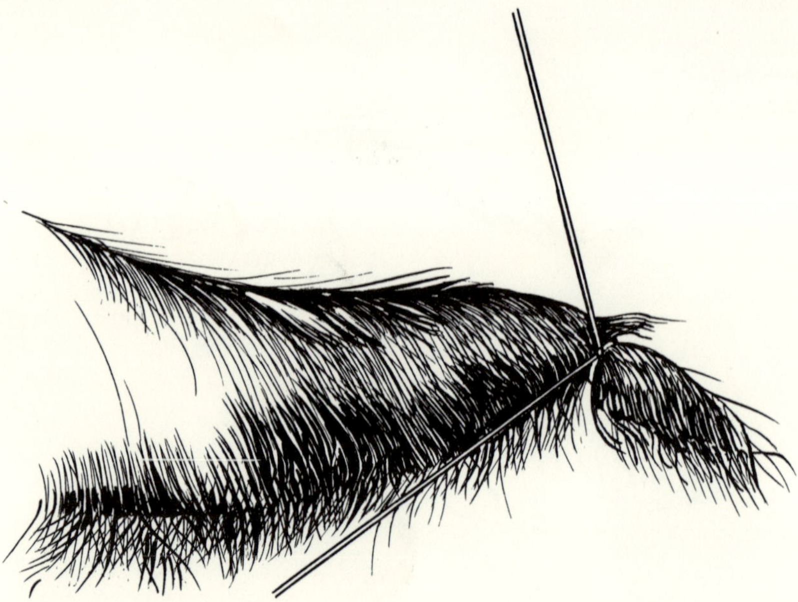

FIGURE 6

outdoor temperatures are high. The best time to perform the operation is during the cool autumn months, which is well before the nesting season for the majority of birds.

The pinioning of very young birds is quite easy and may be done immediately after hatching by simply cutting off the end of a wing, beyond the thumb, with a pair of scissors. Actually there is very little bleeding by this method, but it is almost impossible to determine how close the pinion will be when the bird grows up. Using this procedure a pinioned wing may have an inch or more of bone that should have been removed and from which some flight feathers may grow. By deferring the pinioning until just before the primary feathers start to appear, a short, neat pinion is achieved when the bird reaches maturity. Simply cutting off the end of the wing at this mature age and letting it go at that is not advisable, because there will be an excessive loss of blood. Prior to amputation, a string should be tied tightly around the end of the wing as close as possible to the base of the thumb (Figure 6). This tie-string need not be removed as it will eventually fall off.

At the Philadelphia Zoo well over a thousand birds have been pinioned by the methods just described without a single death or instance of infection.

EDITOR'S NOTE. Recent improvements in tenotomy have been made by James C. Miller, D.V.M. at Patuxent, Maryland. Briefly, Dr. Miller found that by burning through the extensor tendon into the synovial capsule, combined with immobilizing the wing in folded position for several weeks, the birds were permanently pinioned without amputation.

WILLIAM HUEY

Releasing Ducks

The aviculturist has a responsibility to avoid the accidental release of any waterfowl from his collection. In most states it is illegal to make releases without first obtaining a permit from the appropriate regulatory agency. Regulations of this kind are not arrived at arbitrarily but are designed to protect the native wildlife.

Also to be protected is our knowledge of the movement and habits of all wild bird species. Releases, particularly of non-indigenous species, may contribute confusion to this knowledge, and although the ill effect might be obscure, compounded it might eventually result in conflicting and confusing information. Systematic releases of waterfowl from a collection might also have an adverse effect on the native populations, either through competition or by some interruption of established migration, feeding, wintering, or nesting procedures.

There are plus values to the release of waterfowl, and aviculturists have in the past certainly made significant contributions to the status of various species of waterfowl through their efforts. These benefits are primarily in the re-establishment of rare or endangered species, or toward the establishment of some species into a habitat niche which is not currently occupied. Most state departments of game and fish have experience in these areas and could advise aviculturists of possible activities or projects to benefit the perpetuation or restoration of waterfowl in any area. At any rate, no releases should be made without the prior consent of your state game and fish agency.

ROBERT W. GRANT

Shipping

Whenever we want new stock or want to dispose of surplus stock, we are faced with the problem of shipping birds. There are a few breeders who advertise: "Birds must be picked up at the aviary—no shipments made." While this is impractical for universal adaptation, at one time or another we all seriously consider this idea after the loss of some rare specimens during shipment.

There are two means by which shipment can be made: surface (truck, train, or ship) and air. The latter is by far most preferable. Game birds are wild by nature. Being closed up in a crate, then bounced around amid strange noises and odors, is to say the least a traumatic experience for them. Some birds probably do die from fright. It is important then that we subject the birds to this horrible experience for the shortest time possible.

One of the advantages of air shipment is that it gets the birds to their destination in the shortest possible time, where they can be released from their crate and torment. I have been accused of crediting my birds with human feelings. While they do not reason, they certainly can see, hear, smell, and feel. They can become confused and frightened to the point where they never do regain their natural composure, including breeding instincts. The man who has no compassion for the feelings of his birds has no grounds upon which to complain about his losses.

If we must ship by a slow surface carrier, the greatest attention must be given to the crate, and survival conditions, including food and water resupply. Before we discuss these details, let us consider another aspect of the carrier. We have a choice between individual freight organizations and express companies. There are advantages and disadvantages to each. The only national express carrier at the present time is the REA (formerly Railway Express Agency).

A land freight carrier is generally not suitable for small shipments. If you have enough to justify a box car and a special attendant, well and good. But for most of us this is ridiculous. Land LCL (less than car load) shipments are impractical. Water shipments should not be made unless an attendant is to accompany the birds. Air freight is another thing however. If you are shipping to another town served by the same airline, this is the best possible service. You can reserve the required space on a specific flight and also get confirmed space reservations if transfers to connecting flights are required. Other good features of this service include: gentle handling and responsibility assumed by most airline employees; the necessity to confine the bird in a crate for the minimum time before departure; and finally, sure arrival time and flight number, so purchaser can meet the plane and release the bird in his pen without delay. I have shipped birds air freight from San Antonio, Texas, to Mexico City with only a 6-hour lapse from my pen to the purchaser's pen and from San Antonio to New York with an 18-hour lapse from my pen to the purchaser's pen. This not only is easier on the birds, but it also minimizes the need for water and feed. When you are SURE of the time your shipment will arrive you generally need to supply each bird with only one-half apple to satisfy both food and water requirements.

The chief disadvantage in air freight is cost. However, if your shipment is about 50 pounds and on one carrier, it will cost you less than air express. If your shipment must be transferred from one line to another, you will pay at least the minimum tariff on each line. On some long flights this may be acceptable, but on short hops the cost can be excessive. My advice is to learn all you can about the air freight carriers in your area and use them whenever practical.

Due to the demise of many passenger trains over the past few years, live shipment by rail is fast becoming a thing of the past. In the near future there may be no rail shipment of live freight by the REA. Closed and locked trailer truck operations, which are taking the place of the trains, are not suitable for live shipment. Shortly, REA may offer only air service. While this is a blow to the breeder located far from an airport, it is actually in the best interest of the birds.

Another disadvantage of REA shipments is that you do not know exactly when your shipment will leave, how it will travel, or when it will arrive. They are unable to furnish tracer action because they do not keep way-bills on each individual shipment. This lack of control over individual shipments is a considerable handicap to the breeder. He does not know whether his birds will arrive in three hours or three days.

If you have made friends with your local REA man (in a small station) you may receive service fit for a king. However, in large stations the service seems to be degraded in inverse proportion to the size of the station. A shipment may sit around days before it goes out or be lost for several days after it arrives. This of course always seems to happen to the rarest and most expensive birds.

While you pay for insurance (excess valuation) in addition to premium rates for live freight, it frequently is most difficult to collect for losses suffered due to delay or mishandling.

ROBERT W. GRANT

FIGURE A

Old-Fashioned Shipping Crate of Wood. The Top (Inside) of Crate Should Be Padded to Protect Head of Bird. Rough Burlap as Shown Is Abrasive to Tender Heads

On the other hand REA service is convenient and frequently the most economical way to ship birds. Many breeders use it exclusively. However, when making fairly large shipments and when the welfare of the birds is more important than the slight difference in cost, I would recommend air freight contracted directly with one of our airlines.

I have purposely painted a dark picture of the shipping services so that greater attention will be given to the shipping containers. For years many breeders have been shipping birds in wire-bound fruit/vegetable

FIGURE B

Notice "Hold at Airport" Label. It Is Always Best for the Purchaser to Instruct the Shipper to Mark the Crate with the Instructions "Hold at Airport" and to Phone the Addressee at Once. Thus the Shipment Can Be Picked Up at Once, Avoiding the Delays of Local Delivery

crates made of very thin wood slats. These crates are cheap and light. They are easy to fasten and to handle. As far as the breeder is concerned, they are ideal. But not so with the bird!

First, the wood and wire surface is abrasive, making scalping and sore, infected spots on the birds much more likely. They also allow the birds to see out all four sides. In other words the bird has no place to hide. He feels he is surrounded, and he is made more wild by the movements visible on every side. This results in increased jumping and surely adds greatly to the trauma experienced by the bird.

While being open on all sides may permit good ventilation, it frequently submits the bird to strong drafts. Symptoms of resulting lung problems may not show up for several days or even a week after the bird has been placed in the purchaser's pen. Breeders often wonder why a bird, which arrived alive, died only 5 or 10 days after arrival.

The REA has suggested the use of treated cardboard cartons, the type used for shipping apples, lemons, and other fruit. This type carton has a top and bottom or outer and inner section. Each section has 5 sides so that when the top is placed over the bottom, the 4 vertical sides are double walled. The cardboard is treated to resist moderate dampness and unless severely abused will give no problem.

The smooth surface of the cardboard is less abrasive and protects the bird from excessive injury when it makes a violent movement.

Another good feature is that the bird can only see out the ventilation window (which must be cut in the carton) and feels hidden from all the strange sounds and movements around it. A bird is much less likely to move in darkness, as this too gives it a feeling of being hidden.

Finally with only one opening, the bird may be provided ample fresh air without being subjected to drafts.

Large geese, swans, and peafowl require larger crates. Also special crates are required for overseas shipments. But for the average game bird to be shipped by air within the U.S., Canada, or Mexico, the cardboard carton is ideal.

Place one of the molded fruit dividers or several layers of burlap or similar material in the bottom of the carton to give the birds good footing. They will slide and possibly break a leg trying to gain a foothold on paper or similar material.

On top of the molded divider oı burlap place sufficient poultry litter to absorb the droppings and reduce the height of the interior so that the bird can stand *almost* erect (but not quite). This will discourage jumping but allow the bird sufficient movement to be comfortable.

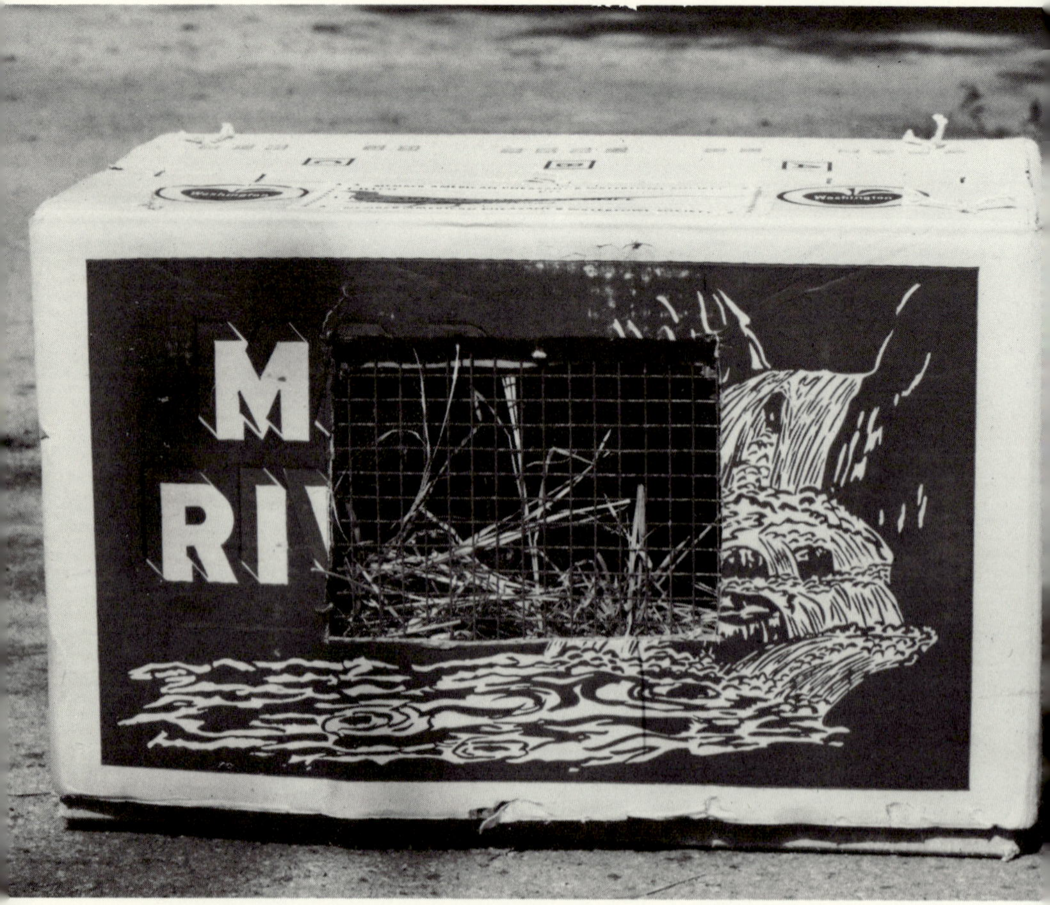

Carton Assembled. Hay Can Be Seen Through Wire-covered Opening. Wire Is Sandwiched Between Two Halves of Carton

In the top section of the box place a piece of foam rubber 1-inch thick to minimize damage to the top of the bird's head. This material is available in department stores in rolls and can be obtained by the running foot. A piece to fit the carton will cost about fifty cents. Using a carpet needle and heavy twine, tack the foam rubber to the top of the carton at four corners and in the center.

On one side of the carton cut a rectangular hole approximately 6 by 8 inches. On the outside of the inner carton attach a piece of hardware cloth (½-inch mesh) with more of the heavy twine, tacking at each corner and in the middle of each side. When the outer half of the carton is placed over this, raw edges of the wire are shielded from both the bird and the carton handler.

Before putting in the bird, add some apple, lettuce, or whatever watery food the bird has been used to eating. It is a good idea to attach the food in a corner by sewing through it and the carton. This keeps it from becoming lost in the litter and it also prevents the bird from soiling the food. One apple (cut in two) or one-quarter head of lettuce per bird should be sufficient.

If over three days are required to transfer the bird from seller's to purchaser's pen/pond, some food and water should be supplied. Fasten either two cans (size #2½ or #303) or the lower 4 or 5 inches of two milk cartons in the back corners. Use a carpet thread loop around the top and bottom of each container and fasten through the side of the inner shipping carton. Directly above the can or milk carton to be used for water, make a 1½-inch hole in the top of shipping carton. Be sure the foam rubber pad is trimmed back so that it will not obstruct access to can. Mark top of carton "ADD *ONE* CUP OF WATER DAILY" and draw or paint a large arrow pointing to the hole.

Just before putting the bird in the shipping carton fill the water container half full and fill the feed container with a mixture of the dry feed the bird is known to have been eating. Do not put feed in bottom of crate for the bird to soil, and do not fill water container so full it will spill out and wet the litter. Always use narrow containers as wide, shallow ones will spill onto the bottom of shipping crate.

Some apple cartons appear to have been treated more heavily than others. If you find you must use a carton with minimum coating, spray (seal) the inside and outside of the bottom (and about 4 inches up each side) with a quick drying enamel or synthetic base paint. This will protect against the softening of the carton bottom by water spilled during the handling, as well as against dampness on the floor of the shipping dock from softening the bottom of the carton.

Crates for larger birds will depend upon the size and weight of the bird. Most geese can be shipped in a standard lettuce crate, while a swan or peacock may require three lettuce crates joined together.

Once the size of the crate has been determined, remove the slats from the sides of the crate and stretch one layer of burlap from the base to the top. Replace the slats, fastening the burlap under the slats as they are re-nailed. Treat the rear (side) of the crate similarly.

The top of the crate should be covered with foam rubber padding or a pad made of sheeting (not burlap) filled with dry hay or straw. Pack the straw filling so that it will be firm enough to prevent ridges in the top of crate from being felt.

The bottom of the crate should be covered with a double layer of burlap on which is placed poultry litter to absorb droppings.

The front of the crate should be made of wood slats spaced close enough to prevent the bird from sticking its head out. Near the bottom cut out the sides of two pair of slats so that a 46 oz. (large fruit juice) can will fit three-quarters inside and one-quarter outside the crate. This is done to facilitate filling the cans from outside. These cans may be fastened top and bottom with bailing wire loops.

Again supply dry foods the bird is known to eat. Fill the feed can half full, also tack a strong cloth sack, containing another half can of feed, to the top of the crate. Mark the slats next to one can "DRY FEED" and the other "WATER." Also mark the sack on top "EXTRA FEED—EMPTY INTO *FEED* CAN ON__(date)__." Make this date about three days from the date the bird is shipped.

Shipping birds overseas is more demanding in crate design and construction. The crate must be strong, small, and clean. The best crate I have seen is built by an exporter in Bangkok, Thailand.

This crate frame is made of wood strips ½ x ½ inch. The sides front and back are 3-ply plywood. The top and bottom are covered with ½-inch wire. The wire bottom is suspended over a removable tray, for droppings, made either of light gauge metal or plywood.

The top has a strong muslin cloth suspended about 2 inches below the wire to protect against damage to the head but to permit fresh air to enter. Each crate is made up of several compartments (usually two to four), each containing one bird. The measurements of each compartment depends on the size of the bird to be shipped. The bird should be able to stand erect (just barely), to roost (squat down) without crowding, but not to be able to turn around. While each compartment is narrow, it must not be so small that the bird is cramped.

The size of the feed and water containers are suited to the bird (see above) and are fastened in openings cut in the plywood so that water and food can be added from the outside.

Until a breeder is very experienced in shipping birds (and importing) it is best to export your birds overseas through a dealer who will construct the proper crates and handle all other aspects of the transaction. This is no game for a beginner.

In summary, ship birds so that they will go from your pen or pond to those of the purchaser by the fastest possible means. You have spent much time and money raising your birds, don't destroy them by being careless, rushed, or just plain stingy in the preparation of the shipping crate. Remember, your birds do not know what is happening to them! Do the best you can to protect them from outside sounds, smells, and sights and from hurting themselves in their natural struggle to escape from the shipping carton. The purchaser has paid good money for live, sound birds, he should not be sent a claim action for dead or crippled birds. Your work and responsibility as a breeder is not completed until your birds have been safely transported to the purchaser.

PART IX

Good Practice

WILLIAM FERGUSON

The Importance of Weighing Ducklings

Shoveler

Assessing the day-to-day condition of ducklings of difficult-to-rear species is one of the hardest observational problems facing those rearing waterfowl. The maintenance of daily weight records is a sound approach to the problem.

Upon hatching, each duckling should be marked for individual identification (spots of artificial food dyes on down and nail polish on beaks are among the many possible methods) and his weight recorded. Thereafter daily weighings will permit the duckling's weight changes to be compared with his previous weight, with those of his siblings, and with the charts provided below to determine if progress is satisfactory or whether special care is necessary. For the results to be consistent it is best to weigh the duckling at the same time each day. A laboratory scale which will register weights to within $\frac{1}{10}$ gram is recommended.

Typically, the weight of a duckling will decline for two or three days following hatching, reflecting a loss of fluids and the use of yolk

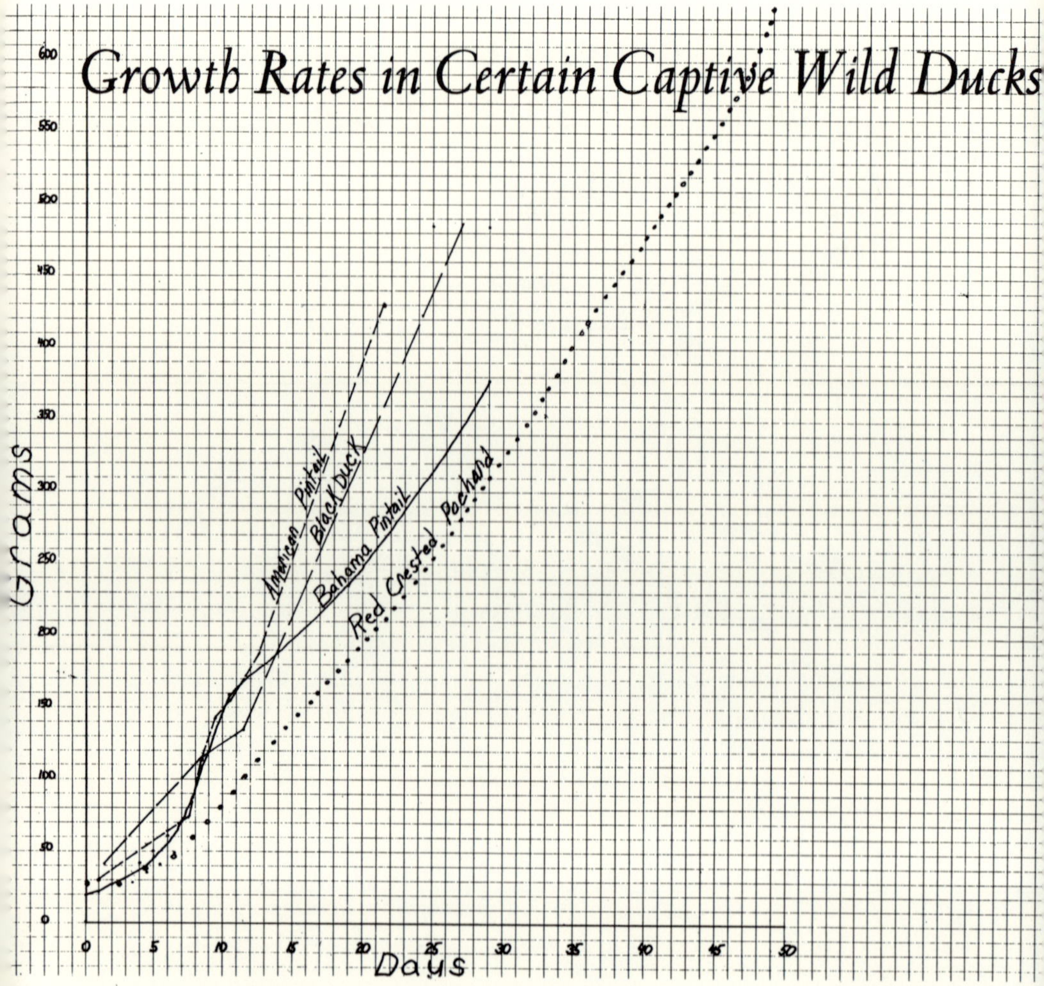

Growth Rates in Certain Captive Wild Ducks WM. FERGUSON

and other energy reserves during a period when the ducklings early feeding attempts are inefficient. Naturally, there are many species-specified weight-gain-patterns. For example, large northern forms tend to grow more rapidly than tropical ducklings. The accompanying graphs illustrate typical age and weight relationships, in hopes that they will give the beginner a point of reference.

For the past three years, I've been weighing newly hatched ducklings from the day of hatching and periodically thereafter through the first few weeks of life.

From such data, the graphs presented here have been extracted. Note that the total number of birds weighed was very small, never being more than 8 of a given species at one weighing and sometimes only one. The points plotted on the graphs represent average weights of birds, sometimes from different hatches and sometimes a year apart as to date of hatch. I made no attempt, for this paper, to separate sexes within a species as to weight. I often found that when several species were placed together in a small brooder, the more voracious ducklings seemed to gain weight rapidly at the expense of the others. When a given group of ducklings grew to a size I judged big enough to transfer to a larger brooder, those remaining in the original smaller brooder shot up in weight promptly.

Certainly, conditions varied a great deal; still, the graphs are tentative suggestions concerning the approximate growth curves of certain species of wild ducks, reared in captivity in central Florida. Similar growth curves utilizing different diets in other climates by other aviculturists may be quite different. Perhaps these curves will stimulate others to publish their data concerning these and other species of wild ducks kept in captivity.

(Please note that all weights are in grams. The scale used was an Ohaus, Triple Beam Balance, 2610-gram capacity. The day of hatching is considered to be day 0.)

BOB ELGAS

Banding

If the aviculturist is to successfully keep track of his birds he will find a banding program to be not only a valuable aid, but indeed a virtual necessity. A well planned banding system, used in conjunction with a good bookkeeping arrangement, will prove its worth repeatedly.

There are a number of commercial sources from which bands (or rings) may be acquired. Bands to suit individual requirements can usually be obtained through merchants dealing in poultry supplies, with varying sizes, styles, and custom lettering and numbering available. If the aviculturist is to derive the most from his banding program he may find it more satisfactory (and less expensive) to make his own bands. This is especially true for those who might not have large flocks. Aluminum is the correct material from which bands are to be made, and most hardware or plumbing and heating establishments have good stocks from which to choose. The material should be heavy enough to be sturdy, yet light enough to be pliable. Soft drawn aluminum, of the type commonly found in roll or flat stock, is far superior to hard drawn aluminum, which is very difficult to work. There are two types of bands in general use—the plain flat band, which is curved upon closure to form a ring, and the strap and eye type which has a positive locking device to prevent loss. Generally speaking the flat band is produced from heavier stock than the lock type since it must rely upon the strength of the material to prevent spreading and subsequent loss of the band. Inasmuch as aviculturists commonly keep birds of varying sizes it naturally follows that an assortment of band sizes will be needed to accommodate large or small legs. Make the band fit the bird—never install a snug band or one that is cumbersome or loose. A correctly installed band is one which is loose enough to ensure free circulation, yet tight enough so that it cannot be slipped over the foot. The band should be loose enough to slide freely up and down the leg. Never install two or more bands on the same leg because they will rub one another and destroy themselves.

The amount of information that can be made available by a banding program is amazing. If the breeder will make a practice of banding all male birds on one leg and all females on the opposite he will be able to correctly identify birds as to sex without the necessity of catching them. With birds which display varying male and female plumage this is relatively unimportant. However, with those varieties in which sexual plumage is identical the value of this system becomes immediately apparent. In ordering (or making) his bands the aviculturist will want his name and at least a minimum address stamped thereon, which will identify any bird carrying his band as belonging to him. Very often an escaped bird is returned undamaged as the result of such an identifying band. For those who might wish to make their own bands, numbering and lettering sets can be obtained through most hardware dealers. With these sets one can stamp any combination of numbers, letters, or combinations of each that he might wish. In my own program I utilize a numbering system which indicates the bird's sex, its age, species, and individual identity. A typical band number, for example, might be as follows—M-651015. The "M" signifies the bird is a male ("f"-for female). Sixty-five indicates the bird was hatched in 1965, while 10 is to identify species. If the breeder has chosen 10 to identify mallards, for example, then all mallards would have 1 and 0 as the third and fourth digits in his number. The 15 (or whatever number is last in the series) would be the identifying number of the particular bird bearing that band. In order for such a banding system to give the best return it should be used in conjunction with the bookkeeping program. Each bird, according to band number, should be catalogued, with all information pertinent to individual birds carefully tabulated. Some of the information one might wish to record, for example, is: source of acquisition; age when acquired; if reared within the breeder's own flock, full information regarding parentage is important; any characteristics or mannerisms peculiar to the individual bird can be charted—in fact the more information the breeder has regarding each bird the more effective the program.

If the aviculturist is faithful in maintaining his banding and bookkeeping system he will soon discover it is a source of highly valuable information. It will help him to determine which birds are good producers, which might be good producers but poor parents. He will better be able to keep track of the ages of his birds and individual characteristics that might be of special importance. With the utilization of a bit of ingenuity a banning program can be made to work very well indeed toward the improvement and stabilization of a high degree of flock productivity.

H. ALBERT HOCHBAUM

Sex and Age Determination of Waterfowl by Cloacal Examination

Sex and age determination of ducks in autumn plumages is a notoriously difficult task. Accurate sex tallies of adults in eclipse are all but impossible except for the expert. Sexes are often alike in juvenal plumage and, after the autumn molt, young birds of many species are indistinguishable from adults.

The difficulties involved in accurate sex identification by plumage characters have been long recognized by ornithologists. One of the oldest axioms is that final judgment of sex must be based upon examination of the gonads (testes or ovaries). But gonadal examination requires the dissection of a dead bird, and dissection cannot be used in large-scale studies of hunters' bags, nor in classifying living birds in banding traps. Nor do the gonads provide an unfailing key to age.

In most mammals, sex can be determined from the external reproductive organs. In all birds, however, these organs are enclosed within a common chamber, the cloaca. In most groups of birds, the reproductive organs in the cloaca are too simple to be easily distinguished. In all ducks, geese, and swans, however, the reproductive organ of the male is a well defined penis held within the cloacal chamber (Owen, 1886). Any duck, goose, or swan that reaches the hand, regardless of age or plumage, may be sexed by the presence or absence of the penis. The penis as a criterion of sex is as constant and as easily recognized as the gonads themselves.

The penis shows two distinct stages of development. In immature males the penis is a small, unsheathed organ; in adults it is large and enclosed within a conspicuous sheath. These two stages may be recognized at a glance. Thus, in the male the penis is a criterion of age as well as of sex.

In the female, a criterion of age is the left oviduct which empties into the cloacal chamber through the left cloacal wall. In immature females the oviduct is closed by a membrane; the left cloacal wall is unbroken. In adult females, the occluding membrane is absent; the oviduct may be seen as a conspicuous slit on the left cloacal well. The occluding membrane never disappears before the young hen attains maturity, hence the presence or absence of the membrane marks the hen as immature or adult.

In both sexes another criterion of age exists in the cloaca, in the bursa of Fabricius. The bursa is a sac-like organ which, in immature birds, extends as a pocket from the dorsal wall of the cloaca. The bursa is always absent in adults. Thus the presence or absence of the bursa distinguishes immature birds from adults.

The unsheathed penis, the occluded oviduct, and the bursa are universal characters of immaturity in all ducks, geese, and swans; but the period of immaturity is not the same in all species. The mallard, for instance, attains maturity and breeds when it is one year old. The Canada goose does not attain maturity and does not breed until it is more than one year old. In the mallard, the cloacal characters of immaturity become adult before the bird is a year old; in the Canada goose the characters of immaturity persist in some or all yearlings. Thus, all mallards showing immature cloacal characters are young-of-the-year; Canada geese showing cloacal characters of immaturity may be young-of-the-year or they may be yearlings.

In all species of ducks studied at the Delta Duck Station,[1] the penis, oviduct, and the bursa become adult in character before the birds are one year old; hence these characters may be used as age criteria only part of the year. During summer, autumn, and early winter, however, these characters may be used to distinguish all young-of-the-year from all adults.

All species of geese probably belong to the group in which some or all of the characters of immaturity persist in yearling birds. Preliminary studies of the Canada goose by Arthur Hawkins[2] indicate that in this species there may be an intermediate stage of development in the penis and an intermediate stage of regression in the bursa characteristic of yearling birds. It is possible that in further studies, criteria may be established by which young-of-the-year may be distinguished from yearlings.

---

[1] Mallard, gadwall, baldpate, green-winged teal, blue-winged teal, shoveller, pintail, redhead, canvas-back, lesser scaup.

[2] Unpublished notes.

Some species of ducks which do not mature until they are more than one year old may, like the geese, retain the cloacal characters of immaturity as yearlings. This group might include the American goldeneye, Barrow's goldeneye, buffle head, old squaw, harlequin duck, and the scoters and eiders.

Few ducks of the above species have reached our hands for study. The limits of use of the cloacal age characters are therefore not known and these species are not considered in the following discussion of the penis, oviduct, and bursa as sex and age criteria.

The duck penis shows two distinct stages of development: the small, unsheathed penis of the immature male and the large, sheathed penis of the adult. These two penis forms distinguished male young-of-the-year from adults.

The penis of the immature male is merely a short, fleshy appendage attached to the forewall of the cloaca just within the cloacal lip (Figure 1D). It shows a conspicuous left-hand twist, giving it the appearance of a miniature pigtail. It is never more than 12 millimeters in length. The penis of the immature bird retains its small, unsheathed form with little or no change in size or appearance until the young bird is 5 to 10 months old. Then, when the post-juvenal molt has been completed, the adult penis is assumed.

In the transition from the immature to the adult penis the organ enlarges rapidly and a new structure, the penis sheath, forms on the left cloacal wall. In its transition from the immature to adult stage, the penis enlarges until it may be 50 to 90 millimeters in length, even longer in some species. The left-hand twist of the immature penis becomes the spiral sperm groove of the adult organ, giving the adult penis a corkscrew-like appearance. The penis sheath, white or gray in color, develops as a conspicuous fold on the left cloacal wall. When the penis and the penis sheath are fully formed, the adult penis becomes enclosed within the sheath from which it protrudes only when distended (Figure 1B, 1C). The penis sheath opens on the forewall of the cloaca, just within the cloacal lip, extending caudally and dorsally along the left cloacal wall where, in dissection, the penis may be seen to pass through the cloacal wall into the body cavity.

These two stages of the penis cannot easily be mistaken one for the other. Except during the very short period of transition, the two stages are entirely dissimilar. With few and minor differences, the two stages show no important variations in different species of ducks. The immature penis of the mallard, for instance, shows no important characters which distinguish it from the canvasback. The adult penis of the redhead is essentially the same as that of the baldpate or pintail. Thus, unlike plumage characters, the penis as an age character is essentially the same in all species.

In examining the penis of a living duck, the bird is held belly up, tail away, with the outer fingers of each hand holding wings and legs. The vent is located with the forefingers. As the vent is located, the tail is pressed backward with the forefingers and the thumbs are placed on either side of the vent. With gentle pressure of the thumbs, separating them slightly at the same time, the vent is opened. As the cloaca is exposed, the short, unsheathed penis of the immature male or the large sheathed penis of the adult will protrude (Figure 1A, 1B). A skilled hand may bring about this protrusion in a few seconds.

In downy ducklings the examination is much the same, although very young birds may be held in the palm of one hand, with thumbs and forefingers of both hands free to make the inspection. Great care must be taken in the examination of newly-hatched ducklings; pressure of the thumbs at the vent may rupture the yolk-sac. At Delta, young captives are not examined until they are two or three weeks old.

In the examination of ducklings, the penis is sometimes held within the folds of the cloaca, not protruding at once. If the penis does not protrude, the forewall of the cloaca should be examined with further manipulation of the thumbs to make certain of the presence or absence of the penis.

In determining the sex of a small goose, the procedure is the same as in ducks, but the bird may be held under the crook of the arm, giving hands more freedom for examination. The examination of a large goose or a swan is a two-man job, one to hold the bird, one to make the examination. In these larger birds it is sometimes necessary to insert a finger into the cloacal chamber to relax the sphincter muscle. The finger should be moistened with mineral oil.

The examination of freshly-killed birds is the same as in living birds, the penis protruding with gentle pressure of the thumbs at the vent. As the vent dries, however, it is necessary to distend it with the fingers or with a probe before the penis may be examined. Examination may be made with greatest ease immediately after killing, while the tissues are still elastic.

There are no limits of time and age in the use of the penis as a sex criterion. Sex determination is possible in downy young of all species, and the presence or absence of the penis continues to be an identification of sex throughout the bird's life.

As an age criterion, the penis is reliable only during the summer, autumn, and early winter. The period of reliability for all individuals of a species closes when the earliest developing young drakes assume the adult penis. With respect to this time limit, the species studied at Delta divide into two groups: Group 1—adult penis never assumed by young males before December 31; gadwall, baldpate, green-winged teal, blue-winged teal, shoveler, redhead, canvasback and lesser scaup. Group 2—adult penis sometimes assumed by young males before December; mallard and pintail.

The species not studied in sufficient numbers to warrant final classification may be grouped as follows: (1) Cinnamon teal, ring-necked duck, greater scaup, ruddy duck; (2) black duck, wood duck.

My evidence for these categories is as follows: Most young captive-reared redheads, canvasbacks, and lesser scaups showed the adult penis in early March; a few showed the adult penis in mid-February, and a few retained the immature penis until late May. Wild birds probably change earlier, but certainly not before January. Hence the penis character is reliable up to December 31.

Most young males among the captive-reared river ducks of Group 1 changed in late January, February, and March; a few not until May. The penis change has never been observed in wild or captive birds of these species before December 31.

A few young mallard drakes began changing in late October at the age of five months; by November 15 they could not be distinguished from adults by the penis. I have observed that a few young wild mallard drakes likewise assume the adult penis by mid-November but most captive-reared drakes and, I believe, most wild drakes do not assume the adult penis until after the end of December. In wild mallards, the penis is reliable until November 15, but after that, and until the end of December, the bursa must be used in aging drakes. After the end of December, the bursa is no longer reliable. In the pintail the same time limits apply as in the mallard. In the black duck the same limits as in the mallard probably apply.

In wood ducks reared at Delta, the adult penis was assumed by young drakes in January and February when these birds were 8 and 9 months old. In the southern United States, however, where some wood ducks nest in late February or early March, we might expect early-hatched drakes to assume the adult penis by late autumn.

In all ducks, geese, and swans, as in most birds, only the left oviduct is functional. The terminal portion of the oviduct empties into the cloaca through the left cloacal wall at a point about midway between the cloacal lip and rectum, or opening of the large intestine. In adult females the opening of the oviduct may be seen as a conspicuous slit on the left cloacal wall. In immature females the oviduct is blocked by a membrane; the left cloacal wall is unbroken. Until the occluding membrane disappears in the earliest-developing young hens, the presence or absence of the membrane marks the hen as young-of-the-year or as adult.

Unfortunately the presence or absence of the occluding membrane is not easily examined in living birds.[3] In dead females, however, inspection of the cloaca is a simple procedure. The lips of the vent are grasped with thumbs and forefingers and stretched to relax the sphincter muscle. With a blunt probe (the quill of a wing feather is always handy and effective) the left wall of the cloaca is examined. In adult females the opening of the oviduct is located ventrally to the large intestine (Figure 1G). The probe may be inserted to verify the presence of the opening. In immature females, the opening is blocked and the oviduct may be examined only by perforating the membrane (Figure 1F).

The period during which the presence or absence of the occluding membrane may be used as an age criterion in all individuals of a species closes when the membrane disappears in the earliest-developing young hens. Because few hens have been available for study in winter and spring I cannot place this date accurately in all species; I can say, merely, that I have never observed the perforated oviduct in young females of any species before December 31. Thus the presence or absence of the occluding membrane during summer, autumn, and throughout December marks all hens as young-of-the-year or as adults.

The persistence of the immature oviduct, however, shows some puzzling variations in the canvasback and lesser scaup. In captive-reared yearling hens of these species, the membrane has persisted as late as January, that is, perforation did not occur in some hens until they were 1½ years old. This persisting membrane in captive yearlings may reflect

[3] If the cloaca is swabbed with a 10 percent solution of cocain, the sphincter muscle of the vent is relaxed sufficiently to permit examination of the cloaca in the living female. Such examination is not recommended in large-scale studies of wild birds. It is feasible only in the laboratory or hatchery.

their slower development resulting from captivity; I have never observed the occluding membrane in wild adults (age judged by the plumage age character of the greater wing coverts). Until further studies have been carried out, however, the oviduct must not be considered entirely reliable in the canvasback and lesser scaup.

The disappearance of the occluding membrane in young females apparently occurs as a normal function of development attending the maturing of the reproductive system; it is entirely independent of any mechanical force such as contact with the penis in copulation or the passage of the first egg. Its disappearance coincides with breeding readiness but is not caused by breeding.

The bursa of Fabricius, described by the noted Italian anatomist Fabricius in 1621, is a thin-walled, sac-like diverticulum which extends as a blind pocket from the dorsal wall of the body cavity between the intestine and the spine. The bursa opens to the cloaca through a duct in the dorsal wall of the chamber, posterior to the opening of the large intestine. As the young bird approaches maturity, the bursa degenerates, finally disappearing entirely.

The bursa is present only in immature birds. It is never present in adults. (This is true of all birds except the ostrich and the rhea; in these species the bursa persists in adult life.) The importance of the bursa as an age criterion in ducks and other game birds has been pointed out by Carl Gower (1939).

Dissection is not necessary in determining the presence or absence of the bursa in dead ducks, but in living ducks, examination of the bursa is difficult.[4] The cloaca of the dead bird is exposed as described in the examination of the oviduct and the posterior dorsal wall is examined with a blunt probe. In immature birds, the opening of the bursa may be found on the dorsal wall posterior to the large intestine (Figure 1D and F). A conspicuous membrane partly separates this posterior portion from the anterior portion into which the intestine empties. The probe may be inserted within the bursa to make certain its presence and to measure its depth.

In adults the bursa and the bursal opening are absent; the dorsal wall of the cloaca is unbroken (Figure 1E and G). All that remains is a small scar in the place of the opening or, rarely, a very small, shallow

---

[4] The bursa of the living bird may be examined if the cloaca is swabbed with a 10 percent solution of cocain. This, however, is a laboratory technique and is not recommended for examination of wild birds in the field.

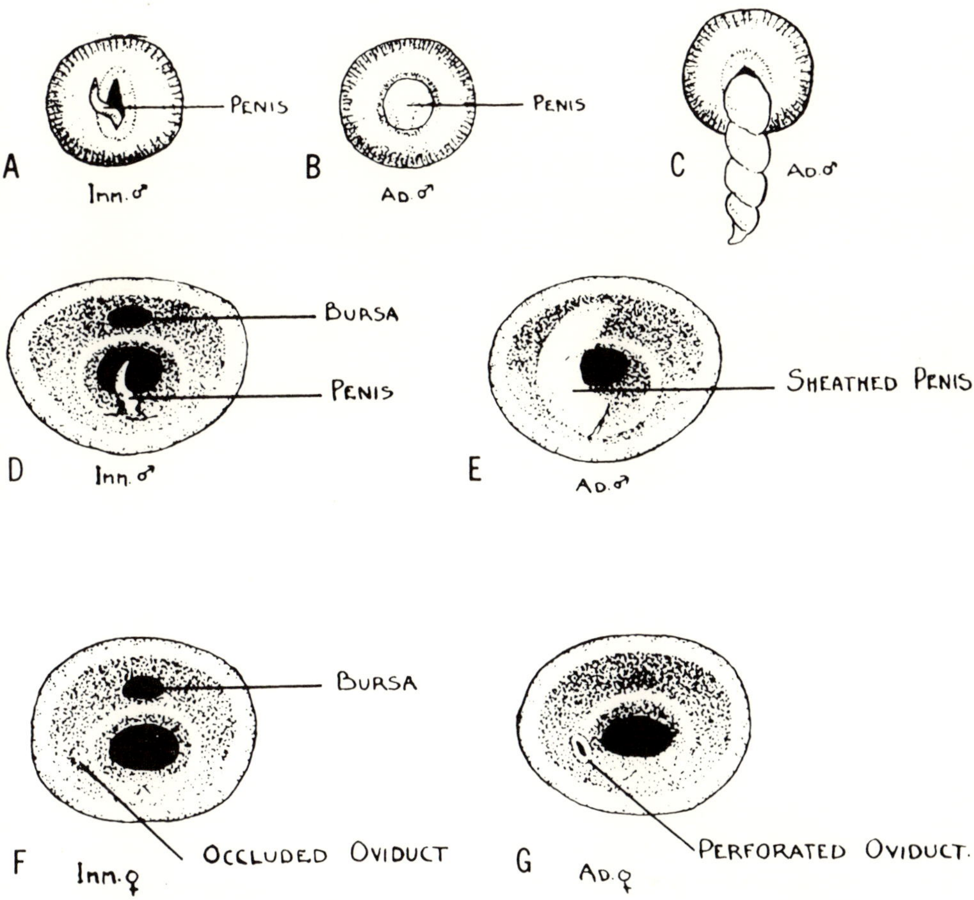

FIGURE I

H. A. HOCHBAUM

Cloacal Characters of Sex and Age
A—Immature penis as seen in examination of living bird. B—Sheathed adult penis as seen in examination of living bird. C—Adult penis protruding from vent. D—Exposed cloaca of immature male showing position of penis and bursa. E—Exposed cloaca of adult male showing sheathed penis on left cloacal wall. F—Exposed cloaca of immature female showing position of occluded oviduct and the bursa. G—Exposed cloaca of adult female showing position of perforated oviduct.

cavity. Eventually in adults the scar and even the membrane separating the anterior from the posterior portions of the chamber disappear.

The bursa attains its greatest size when the young bird is two to four months of age. The maximum development is attained in river ducks at an earlier age than in diving ducks.

At about the time of the post-juvenal molt, a gradual degeneration

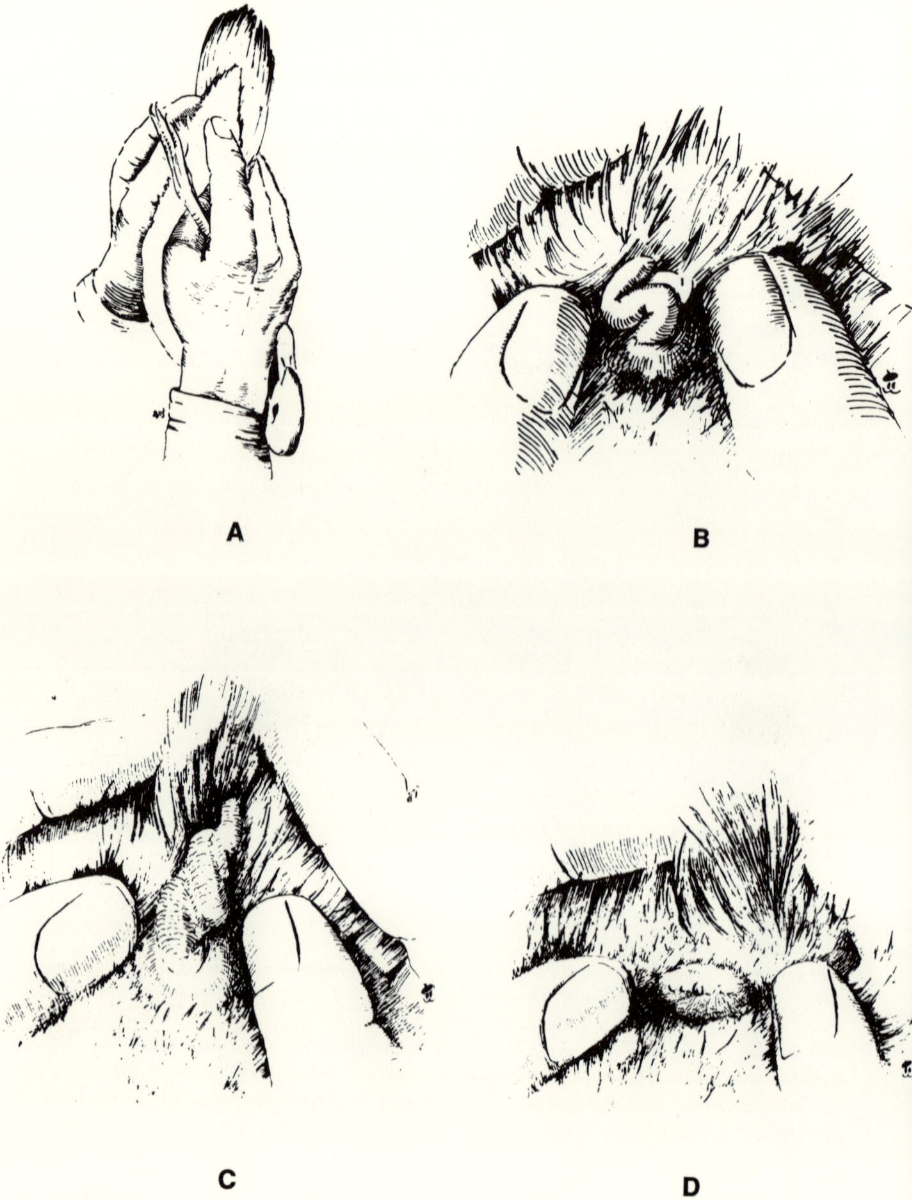

FIGURE 2

RON JENKINS

Sexing

A—Correct method of holding duck for vent sexing procedure. B—Sex organs of immature drake. C—Sex organs of adult drake. D—Sex organs of duck (female)—the raised portion toward the front center is the entrance to the oviduct.

of the bursa begins; it becomes shorter and its cavity smaller, until it finally disappears.

The period when the bursa may be relied upon as an age criterion in all individuals of a species closes when the bursa disappears in the earliest-developing young. Early-hatched pintails, for instance, may lose the bursa by the middle of January, after which such birds cannot be distinguished from adults by the bursa character. Other pintails, presumably late-hatched or otherwise retarded, retain the bursa until April or May. But after the early-hatched birds have lost the bursa, it is no longer reliable as an age criterion for all individuals of the species.

It has not been possible to examine many ducks in winter and spring, hence I cannot date accurately the earliest disappearance of the bursa in the various species. I have never observed complete degeneration in any species before the end of December. Carl Gower, in his Michigan studies, likewise found the bursa always present in young ducks throughout December. Pending further studies in winter and spring, I must consider the end of December as the close of the period when all individuals of a species may be aged by the bursa.

There is much room for further study of these cloacal characters. All time limits for their use need to be refined species by species. We do not know when the transition from the immature to adult penis occurs in geese, swans, and many species of ducks; nor do we know when perforation of the oviduct and complete degeneration of the bursa occurs in many species.

In this present study, nevertheless, we have examined the cloacal characters of more than 5000 ducks in hunters' bags, banding traps, and the Delta pens. In addition to this, Arthur Hawkins and other field workers of the Illinois Natural History Survey have studied these cloacal characters in several thousand ducks on the Illinois River. Both the Delta and Illinois studies indicate that within the defined periods and with the species studied, the cloacal characters are reliable.

I need speak but briefly of the importance of the cloacal characters of sex and age. They greatly expand the possibilities of autumn bag tallies. Already the Delta Duck Station and the Illinois Natural History Survey have a 4-year record of the sex and age composition of hunters' bags in Manitoba and Illinois. The Delta studies have shown, among other things, that on breeding-ground shooting areas, the kill of adult females in most species is much heavier than the kill of adult males; that the kill of young females in some species is heavier than the kill of young males; that differential sex and age migrations and different types of shooting play an important role in the sex and age make-up of hunters' bags.

We are giving more and more attention each year to our waterfowl crop; counting our stock in winter and summer, improving wintering and breeding ground environments, appraising yearly production, but we know very little regarding nature of the legal harvest. Samples of the yearly loss to guns should be studied not only in Illinois and Manitoba, but in many parts of the shooting range.

Bibliography

Gower, W. Carl, The use of the bursa of Fabricius as an indication of age in game birds. Trans. 4th N. Am. Wildlife Conf., 1939, pp. 426–430.

Owen, Richard, On the anatomy of vertebrates. London: Longmann, Gunn & Co., 1886. Vol. 2, p. 244.

WILLIAM HUEY

Attracting Ducks

Red-Breasted Merganser

I am not convinced of the desirability of attracting waterfowl to the premises of an aviculturist, although I am sure it would be considered highly desirable by most aviculturists to have visits from those species which are rare in captivity, or at least not in his own collection.

It is a great deal more likely that the waterfowl attracted to the private collection will be the more common forms, probably mallards. Although there are few sights more appealing than a flock of mallards over a wild marsh, a flock of mallards in a private collection can become quite a nuisance. In the first place, they will compete quite voraciously for feed, and their continued presence will probably require special feeding techniques to avoid a substantial increase in the feed bill. There is, of course, also the ever present danger that these wild birds might introduce disease or parasites into the captive flock. The aviculturist also has the responsibility of avoiding the possibility that his captive birds, or wild birds attracted by his captive birds, might have a decoying effect on some hunting area. This, of course, is illegal and represents one of the most serious reservations that the Federal Regulatory Agency has toward the private practice of aviculture. These negative aspects should certainly be given serious consideration before the attraction of wild birds is considered.

If you still want to attract wild ducks, consider these steps. Essential to the attraction of wild birds, of course, are two factors—feed and water.

Although some species of ducks are quite commonly attracted to good feed conditions in dry fields, the most critical of the two attractive features is water. Those of you who fly know that a pond of water is visible for a considerable distance from the air, and the reflection of light on a pond is practically irresistible to a duck that is looking either for a place to rest or something to eat. The type of water and the abundance of feed present there will govern whether or not the birds stay to attract other birds or whether they move on. I would say, generally, ducks prefer shallower water for feeding, with sand bars, islands, logs, or gently sloping banks as loafing spots. These loafing spots are not so attractive to divers as to the puddle ducks or river ducks, since the loafing of these diving species usually occurs in deeper, open water.

It is difficult at best to maintain any natural food in ponds used for the captive flock. Where pressure is extremely heavy, clumps of cattail or bulrush may be maintained within protective enclosures of 1″ mesh poultry wire. This provides little more than a landscaping effect, however, as the food production from plantings of this kind is minimal. This means that feeding must be largely artificial. All ducks, of course, are attracted by corn and small grains scattered in the water. Almost all species may be coaxed onto the banks for feed, whether on the ground or from feeders, and the wild ducks are no exception to this rule.

Those aviculturists who are located along established routes of waterfowl travel will find more visitors on their pond than will the aviculturist located off the established waterfowl routes. A pond with open banks will probably attract more waterfowl than one with big trees and overhanging vegetation on the perimeters, since the open pond would be visible for greater distances.

A gimmick, which is certainly effective during heavy winters, is a system for ice melting which will provide open water in areas when all other ponds are frozen. This is easily accomplished with an oilless air compressor pumping air into perforated plastic tubing laid on the bottom of the pond. It is possible to control the operation of the pump thermostatically so that it only operates during those periods when it is necessary. This ice-melting system also provides comfort for captive waterfowl during those periods when some protection would be necessary if provisions for open water were not made.

Again, however, one should carefully consider the complications of attracting wild waterfowl before you decide whether you want to attract wild waterfowl to your area.

PAUL A. JOHNSGARD

Photographing Ducks

The subject of bird photography is a most difficult one to cover in the space available here, and the interested reader is advised to read such books as *The Technique of Photographing Birds* by John Warham, or Eric Hosking's *The Art of Bird Photography*. A succinct and authoritative account may also be found in *The World of Birds* by James Fisher and R. T. Peterson. The present article will assume a basic knowledge of such things as exposure and fundamental techniques and will concentrate on the special problems of photographing waterfowl, either in captivity or in the wild. Motion-picture photography is a special problem and will not be considered here.

Cameras and Lenses

Most amateur and professional bird photographers now use 35 mm. cameras, and this discussion will center around such equipment. Without doubt, the most suitable camera for a person interested in bird photography is a single-lens 35 mm. reflex with available interchangeable lenses, including long focal-length lenses. Such lenses should preferably have automatic diaphragm provisions which will "step-down" the aperture to a pre-selected opening at the time the shutter is activated. Secondly, the camera should have either fully automatic exposure provision (perhaps most suitable for those persons who do not trust their own abilities to manually control exposures) or have a light meter that "reads" directly through the lens. Two fully automatic exposure single lens reflexes that have available lenses of considerable focal-length are the Topcon Auto-100 and the Voigtander Ultramatic; for persons who do not desire manual exposure control, such a camera might be a very good choice.

A larger number of brands are available that lack fully automatic exposure control but do have through-the-lens exposure readings coupled to shutter and film speeds, and a variety of automatic diaphragm telephoto lenses available. These include such well known brands as the Nikon and Nikkormat, the Pentax Spotmatic, and Topcon Super D and DI, to name only the most widely known and top quality brands. Any of these would make an excellent choice, with the decision on a specific camera dependent on personal preference concerning variations in lens mounts, focusing screens, and available extra attachments such as motor drives or specialized lenses, or simply handling and viewing ease. Although long focal-length lenses are obviously necessary for photography of either wild or captive birds, it is rare that the waterfowl photographer will need a lens in excess of 300 mm. Lenses of 400 mm. or longer are usually excessively heavy, normally require tripods or other supports, and the optical difficulties associated with maintaining lens quality in manufacturing such long lenses force their prices to be relatively much higher than those of shorter lengths. Thus, a 600 mm. lens would have a magnifying power only double that of a 300 mm. lens, but in both weight and price is likely to be at least four times as great. For the rare instances when a photographer needs such extreme magnification, a fairly inexpensive 2X converter attached to one's 300 mm. lens will usually provide a usable if not professionally perfect image. Many excellent 300 mm. automatic lenses are now available with f/5.6 maximum apertures and weights of less than 25 ounces. Such lenses not only are adequate for photographing most wild waterfowl but, if they can focus to within 10 feet or so, will also serve as excellent portrait lenses for captive waterfowl.

Otherwise, a 135 mm. or 200 mm. lens is very useful for close-ups of captive birds which can be approached to within 5 or 10 feet. In general, a combination of a 135 mm. and a 300 mm. lens should be adequate for nearly any waterfowl photographer's needs. Few zoom lenses adequately cover this entire range of focal lengths, and those that do are not only heavier and more expensive than the corresponding fixed-focal length lenses, but also tend to be optically inferior to them. Since the lens quality places an ultimate limit to photographic quality, it is better to buy a single, top quality telephoto lens and rely on it, rather than to get a variety of lenses having various focal lengths but uniformly mediocre quality.

Films

Most amateur waterfowl photographers will probably want to use color film exclusively, and several excellent films are available that may fill their needs. Important criteria include color fidelity, film speed, contrast and shadow detail, and "graininess." No film is perfect in all respects; as film speed is increased it is usually true that image quality deteriorates, and both contrast and color characteristics may become objectionable. Persons wishing to photograph flying waterfowl and thus needing shutter speeds of 1/500 second or less will be forced to use the fastest available films and to accept their color, contrast and grain characteristics. Of those "fast" films currently available, the High Speed Ektachrome (Daylight) offers the best combination of qualities in the opinion of the writer, but those who can tolerate the greater contrast and graininess of Anscochrome 500 may be willing to compromise for the sake of a higher film speed.

A much wider range of films are available in slower speeds, which often are entirely adequate for photographing captive waterfowl under somewhat controlled conditions. Here personal preferences in color rendition will allow for greater diversity of opinion, but some films which in the writer's opinion are particularly useful for bird photography are Agfachrome and Ektachrome-X. These offer moderate film speed, great image detail, brilliant colors, reasonable contrast characteristics, and uniformly high quality control in production and processing. Although some persons may wish to carry two or more brands of such all-around color film in their camera cases for special uses (Ektachrome is excellent for sunsets; Agfachrome produces beautiful skin tones), the average photographer should learn to use a single color film very thoroughly and be aware of its strong points and limitations, rather than unselectively obtain whatever happens to be available at the time.

Black and white films are also available in wide variety, and the choice of a single film is largely dependent on the compromise the photographer is willing to make between maximum film speed and minimum graininess. A convenient film speed for black and white film is 500; this allows the photographer to use 1/500 second as a basic shutter speed, with a general rule-of-thumb aperture setting of f/16 for average subjects in sunlight and increasingly larger aperture stops for cloudy-bright, cloudy-dull, and open shade conditions. Subjects that are mostly white, such as most swans and snow geese, should be given less exposure than normal subjects to avoid losing feather details through overexposure, although this is more critical for color film than for black and white.

Techniques

Techniques used for photographing captive waterfowl obviously differ from those necessary for photographing wild birds, the latter being much more difficult and demanding of time. With captive birds, manipulation and prior planning are often possible that enhance a photo's value considerably. Sun-angle can be chosen (usually front-lighting from an angle is better than is light from directly overhead or back-lighting), and mid-morning or mid-afternoon is usually ideal. For color photography of birds with delicate coloring, a slight overcast, or otherwise working under conditions other than direct sunlight, will often help to enhance soft colors. However, iridescent coloration requires direct sunlight and a fairly specific sun angle in order to produce the desired effect. Also, unless a flash attachment is used, direct sunlight is needed to produce the "highlight" in a bird's eye that helps to add life to close-up photos.

Photographing waterfowl on water requires consideration of the effect of light reflections on exposure and color rendition. Even very dirty water can be made to appear a beautiful blue by shooting on a calm and cloudless day with the camera held at a low angle to the water. Lowering the camera angle also helps to give the impression of a "duck's-eye" view of another waterfowl, and avoids the common "penned-in" appearance of many waterfowl photographs. However, it adds to the background problem, and special care must be taken to exclude fences, other birds, or distracting elements by being certain that they will at least be out of focus and inconspicuous.

Unless a "flock-shot" is desired for specific reasons, it is better to concentrate on photographing single birds, pairs, or families. Reducing the number of subjects in a picture increases the chances of finding the proper moment when the composition is ideal. If the birds are pinioned, it is also easier to make certain that the birds are being photographed from the unpinioned side if only one or two birds are part of the scene. When one of the subjects is moving objectionably, or perhaps is asleep, a short whistle similar to the warning note of a songbird will often cause the subject to become momentarily alert long enough for a picture to be snapped.

Flight photographs of wild birds pose several problems, and the photographer must be willing to accept a considerable percentage of wasted shots. Although large geese can often be photographed successfully with a shutter speed of 1/250 second, and small ducks may demand a speed of 1/1000 second, the writer usually has standardized on a speed of 1/500 second and has simply adjusted the aperture to fit the light conditions. Focusing on flying birds is easiest when they are moving at approximately right angles to the camera, and is most difficult when they are directly approaching or flying away. It is generally best to try focusing a single bird in a flock, and to try concentrating on some small detail, such as the bird's eye, until the proper range is reached. Such "flight-shooting" is fairly easily done with a 300 mm. automatic telephoto, especially if the camera is mounted on a "gun-stock" mount that is adjusted to the individual. Although tripods have limited value for photographing birds in flight, they are very useful for photographing standing or swimming birds at great distances, when very long lenses are needed. Under these conditions, heat waves, atmospheric haze, and lens limitations often combine to make waterfowl photography extremely difficult.

Waterfowl photography has a number of valuable benefits, such as recording various plumages, postures, behavior patterns, and the accurate rendition of such features as the bill and leg coloration. These photos may thus have considerable value for artists or for others who wish to obtain accurate information for species in which they are interested. Additionally, it is a pleasant and often challenging way of increasing one's enjoyment from captive and wild waterfowl.

PAUL A. JOHNSGARD

Scientific Value of a Waterfowl Collection

The presently relatively advanced state of knowledge that has developed concerning the family Anatidae is at least in significant portion a result of this group's adaptability to captivity, and such persons as John C. Phillips and Konrad Lorenz obviously utilized their experiences with personal waterfowl collections in developing their important contributions to ornithology. More recently, the enormous value of the Wildfowl Trust's collection at Slimbridge has been proven by the large number of research papers that have been written as a result of the Trust's unrivaled collection.

Although the average amateur aviculturist might feel unqualified to attempt undertaking a major research project, he can unquestionably contribute data that is difficult or impossible for the field biologist to accurately obtain. Obvious examples for a particular species include length of incubation periods, frequency of laying of successive eggs in a clutch, age at first breeding, maximum longevity, and the relative role of the male in nesting and brooding of the young. There is a remarkable amount of contradictory information regarding such things as incubation periods and probable age of initial breeding in the technical literature, much of which has been cited time after time without any attempt at confirmation. A small amount of differences in such figures is to be expected but certainly not the degree of variation which presently exists. Likewise, although information on fledging periods obtained in captivity might differ somewhat from that for wild birds, it does not make this information any the less valuable. For the person who regularly breeds a number of closely related species, a valuable project would be to obtain comparative breeding data for such species when hatched and reared under identical conditions.

Many aviculturists may be tape-recording enthusiasts, and with the presently available portable cassette-model recorders that are exceedingly easy to operate, there are numerous opportunities for recording waterfowl vocalizations. This might include such things as the corresponding vocalizations of a group of related species (such as the "headthrow" call of male pochards, the "descrescendo call" of female dabbling ducks, or the "triumph ceremonies" of geese or swans). Such recordings would be of great potential interest to biologists and additionally might provide amusing listening opportunities.

Likewise, amateur motion-picture equipment of either 8 mm. or Super 8 size is potentially adequate for filming behavior patterns in captive waterfowl. By using 8 mm. film that was loaned to me by various persons, I have been able to analyze the visual displays of species such as the Maccoa duck almost as effectively as if the birds were studied personally. The guidelines here are to film in slow motion (24 to 48 frames per second), to try to film the same behavior patterns several times in order to establish what variations occur, and to try to film all the major aspects of display activity. Unfortunately, copulatory behavior patterns of many species have as yet never been photographed or described in print, although I have been told by some aviculturists that they have observed such behavior on many occasions, but simply cannot recall the details. A survey of the species accounts in my *Handbook of Waterfowl Behavior* (Cornell University Press, 1965) will allow the reader to quickly determine what behavior patterns have been described for any species of Anatidae, and thus evaluate his potential for contribution. Not only rare and exotic species need such descriptions; there are not presently available accounts of copulatory behavior in such familiar species as the European wigeon, falcated duck, Baikal teal, red-billed pintail, garganey, or cinnamon teal, to mention only some of the more commonly encountered dabbling ducks.

The means by which pair-formation in waterfowl occurs is by no means clear in many groups, particularly in swans, geese, and whistling ducks. We need to know at what age this occurs, how long it requires, and what influence individual contacts established during the juvenile stages play in promoting the formation of pair-bonds between individuals. Aviculturists who band their birds, or who otherwise recognize them well enough to individually identify them, might easily help to solve these problems, which have important implications in establishing game management and conservation policies.

Persons who rear species of waterfowl that are infrequently bred in captivity have special opportunities to contribute valuable information. For example, plumage changes in the sea ducks such as eiders and scoters are extremely complicated, and the variations encountered in looking at a series of skins of wild birds are impossibly difficult to interpret. A set of photographs of marked individuals, taken at 3- or 4-month intervals, from the juvenal stages to the adult plumages, would have great value for the interpretations of plumage changes, and might allow the accurate field-aging of immature birds by biologists. The establishment of accurate age-ratios during the fall and winter has acquired great significance in the annual estimations of waterfowl productivity, and these are dependent on the recognition of first-year birds in the wild populations.

Although the death of any valuable bird is a bitter blow to the aviculturist, and he might be inclined to bury his disappointment as soon as possible, he would be far wiser to inquire whether the specimen might have any value to any museums, colleges, or universities that may be nearby. Virtually every portion of a carcass is of value to some biologist, and many aspects of anatomy, such as skeletons or preserved soft-parts, are relatively difficult to locate even in large museums. While at the Wildfowl Trust I began methodically saving every waterfowl trachea that I was able to remove during the regular post mortem examinations of dead waterfowl, and within a year had developed the world's most comprehensive collection of these structures, which have important taxonomic value. Happily, the collection was continued and expanded after I left Slimbridge, and so far at least two major technical papers have been published as a direct result of this "garbage pail" collection. Similarly, the Trust has made a special attempt to save the skins of downy young waterfowl, which also have special taxonomic significance, and as a result of hatching mortality have accumulated a unique comparative series of downy plumages that are beyond value, in spite of the fact that each one represents an avicultural disappointment.

PART X

Endangered Waterfowl

PHILIP STANTON

The Care of Oiled Water Birds

Goosander

Until 1967 almost nothing was known about how to take care of oil-soaked waterfowl. Then came the Torrey Canyon disaster which killed countless thousands—an estimated 20,000 guillemots alone. British ornithologists managed to rescue from the sea more than 6000 survivors. Although they worked in teams, were helped by scores of volunteers, and followed what seemed like the best procedures, only 350 birds were alive six weeks later. And only a very small fraction of these ever returned to a natural existence.

A number of techniques, especially in the areas of feeding and cleaning, have been tried since then. My method, which in many details goes contrary to earlier practices, is recommended here because it works. In contrast to the Torrey Canyon survival rate, almost 90 percent of the birds entrusted to my care have survived; several species such as scoters and loons have never before been treated successfully in captivity. It is a no-nonsense method, designed to handle large numbers of birds with a minimum of manpower, equipment, and expense, and it is simple enough to be followed by the kind-hearted beachcomber who picks up a single oiled bird.

Transportation

We moved oiled birds from the shore to our rehabilitation center in vegetable crates which are used for trucking lettuce and other vegetables. The crates can hold 5 or 6 birds, they stack easily, and the wire mesh allows plenty of ventilation. For just 1 or 2 birds, cardboard cartons with holes punched in the sides are adequate. Loons should be segregated because they are aggressive even when they are covered with oil and close to death.

Pioneer rescuers recommended transporting oiled birds in cloth bags with their heads protruding to keep them from preening and pecking oil. As a rule birds preen only when relaxed, and I have never observed them preening while in transit. However, if we hold them for some hours before cleaning, we tape each bill together.

Cleaning

Many cleaning techniques, often used in combination, have been tried: dusting the oiled feathers with sawdust, fuller's earth, or even cat litter; coating them with castor oil or salad oil; washing with shampoos, household detergents, or special chemicals. I am convinced that most of these procedures are inefficient or worse and so traumatic for the birds that they have caused many deaths.

The cleaning agent that we and other wildlife experts rate as by far the best now available is Polycomplex A-11, a chemical developed a few years ago to disperse oil spills. It gets the oil off fast, without causing much stress to the birds, and has almost no harmful aftereffects. It is distributed by Ocean Wide Industry, Inc., 50 Elm Street, Huntington, New York 11743, costs about $4 per gallon, and is sold in 5-gallon pails. We have compared its performance with that of similar chemicals and find it far superior.

Waterfowl oiled by winter spills in a cold climate should be cleaned in a room where the temperature is not above 60 degrees F. Although rescuers tend to want to make the half-frozen birds warm and comfortable without delay, heat seems to aggravate the frostbitten areas.

We set up at least 2 and preferably 3 tubs, one for washing and the other two for rinsing. Old-fashioned, galvanized metal wash tubs are the correct size and easy to empty and refill as needed.

The standard washing solution is one part Polycomplex A-11 to 100 parts water. The rinse tubs contain water only, and the water in all is at room temperature.

Handling the birds gently but firmly, we dip them in the solution and swish and work the feathers until they are clean. If only a portion of the feathers have been oiled, only that section should be cleaned. Then we immerse the birds in the rinse tubs and pat them dry with old cloths or towels.

Holding Bird Prior to Washing MRS. ROBERT T. PAINE

Washing Away Oil

Bathing

Rinsing PHOTOS: MRS. ROBERT T. PAINE

Polycomplex A-11 is so efficient that 2 or 3 birds can be cleaned in 2 or 3 minutes and with one washing. A few may need a second washing, which is always done immediately.

Shelter

The birds are now moved to a dry, roomy enclosure such as a chicken house where they will spend the next few weeks recovering.

We keep the temperature at about 60 degrees in the winter and as near 70 degrees as possible in the summer.

We cover the floor with 4 inches of serval (crushed sugar cane). Since most sea birds are not physiologically adapted to either resting or walking on land, they need this deep, soft cushion to prevent the development of breast sores and tumors and to protect their feet. Serval reduces the incidence of aspergillosis, a pulmonary infection commonly brought in on hay or straw litter.

Under severe cold conditions, I add one heat lamp for every 8 birds. The hot spot under the lamp should not exceed 90 degrees F., and the birds must be able to leave the heat at will. For the first few days they seek the heat constantly, then slowly lose interest. When they no longer use the lamp it may be removed.

I allow plenty of room. Even in these unnatural surroundings, the birds set up little territories and in the resting position will peck everything within reach of their bills.

In dealing with a variety of waterfowl, I separate them by species. If all are penned up together, the larger will try to dominate the smaller—eiders will trample and peck scoters, and scoters will harass scaup. However, scaup and goldeneyes may be kept together.

Water

The only water the birds should have access to for the first two weeks is drinking water because they are still weak and prone to chilling if wet. The drinking device should be covered with a grate so that the birds can drink but cannot become immersed.

They drink fresh water readily and with no harmful effects, and later fresh water may be allowed for swimming.

After two weeks, shallow bath pans or ramped swimming pools no more than 18 inches deep must be provided. A child's plastic wading pool will provide swimming space for 6 or 7 birds. All pools must have ramps so that the birds can walk in and out easily; otherwise, without oil on their feathers, they are sure to drown. Ramps can be made by nailing slats on wide boards.

It is a poor idea to sink the pool into the ground for the landscaping effect; in a few hours birds foul their swimming water and sunken pools are hard to drain.

Food

Because we were determined to develop a practical method for handling large numbers of birds, we did no forced feeding. We found that if the proper set-up is provided and the birds are not disturbed for the first 3 days, they will begin eating on their own. We use no vitamin tablets, sugar and water mixtures, or other food additives but provide grit and oyster shells mixed with the food.

The food must be kept dry. If it is wet, the birds smear it over their heads and breasts and are soon in as poor condition as when they were covered by oil.

Recommended Diets

Eiders, American and red-breasted mergansers, and old squaws prefer Gaines dog meal (pelleted form), trout chow (Purina Developer), and game-bird chow (pelleted form), in that order. After a few months in captivity they will eat almost any type of game-bird food.

Scoters and buffleheads choose trout chow (Purina Developer). Gaines dog meal (pelleted form) is also acceptable. They like soaked dog biscuit and lettuce, but it mats their breast feathers and is very messy so should be avoided.

Scaup and goldeneye get only trout chow at first. Later this may be supplemented with game-bird chows and grains.

Dabbling ducks (blacks, mallards, pintails, teals, etc.) get game bird chows, green feed, corn, and other grains.

Gannets, gulls, and terns get fish, fish byproducts, such as entrails from a fish cleaning plant, shellfish, and canned dog food which is moist and meaty in flavor. Avoid canned fish or other food containing additives.

For loons, grebes, alcids (auks, murres, guillemots) one starts with chopped liver, fresh shellfish taken from the shell, or freshly killed small fish such as smelt, chubs, pond shiners, and small yellow perch. Later they may be fed gourmet-type fishcakes covered with crumbs, and finally trout chow and pelleted Gaines dog food.

Medications

During their recovery, oiled birds are commonly subjected to two types of illness. The symptoms should be watched for and treated promptly.

1. Clouded or milky eyes, listlessness, lightness of body, and lack of appetite.

Administer a 50 to 125 mg. dose of terramycin every twelve hours until the condition improves. The size of the dose depends on the size of the bird; dovekies or teal require 125 mg. Watch the droppings, since overdoses usually cause diarrhea. Terramycin can be bought from most poultry supply houses or from a veterinarian.

2. Lameness, dry, cracked legs and feet. A natural consequence of keeping water birds in dry surroundings.

Use A and D Ointment (White Laboratories Inc. Order Dept. 1011 Morris Ave. Union, New Jersey 07083) or Bag Balm (Dairy Association Co. Inc., Lyndonville, Vt.). Treat until normal condition has been restored. These ointments also seem to help frostbitten areas on the legs and feet. Vaseline is an acceptable substitute.

Extended Care

We keep our birds under hospital conditions for 2 weeks. Our outdoor pens are well shaded, each has a tub or pool for swimming, and the species are segregated. Then, as they seem ready for it, we give them more space, more water, and get them outdoors in large pens where they will stay until they can be released. The waiting period may last for 6 months or more.

Polycomplex A-11, like all similar products known today, removes natural oils as well as crude oil from the feathers, leaves the birds without insulation or waterproofing, and seemingly alters feather structure. Consequently, even after they have regained their strength in other respects, they must be held in a protected situation until the natural oils have built up again. Birds which were lightly oiled and cleaned selectively may be ready for release 6 or 8 weeks after the oiling, but most will have to wait for the next complete molt.

Release

Like everything else about the treatment of oiled birds, the final step—their release—sounds simple but in fact must be handled with great care. When they are tossed into the wind or thrown into the water, their chances for survival are slim.

The birds should be taken to a secluded 3- or 4-acre pond with an island. If any are divers, the pond must be within close flying distance of the ocean. Natural salt water or brackish ponds are the best. Artificial ponds, especially those being used by domestic or feral ducks, should be avoided—the water is apt to be polluted and natural food scarce and the wild waterfowl will be dominated by the others. A few of the released birds fly off heedlessly, but most linger around for 2 or 3 weeks, gradually adjusting to the natural environment until they finally disappear.

The procedure outlined here is sure to be modified and improved. My own research is now devoted to analyzing the chemical composition of the natural oils in the feathers of waterfowl. When this is known, either a selective cleaning agent or a synthetic substitute for natural oil can be developed and oiled birds released shortly after they have been picked up.

S. DILLON RIPLEY

Endangered Waterfowl of the World

Steller's Eider

In the early 1960s the International Council for Bird Preservation joined with the International Union for the Conservation of Nature in a project to produce Red Data Books to be published from the IUCN headquarters at Morges, Switzerland. The Bird Volume, Vol. 2, is compiled and kept up to date by Lt. Col. Jack Vincent, Liaison Officer of the International Council for Bird Preservation whose headquarters are at P.O. Box 662, Pietermaritzburg, Natal, Republic of South Africa. This volume lists the species of birds including waterfowl thought to be seriously endangered today, represented by pink colored sheets of information. Species formerly thought to be in danger, but which have recovered for various reasons, are represented by green sheets, and those which are either rare (perhaps in a limited habitat) or which are thought to be depleted and whose numbers are continuing to decline, are listed on white sheets. The volume is looseleaf so that animal species can be added or deleted from time to time.

Most of the species which I list herewith are thus found in the Red Data Book, Vol. 2. However, I include three species not listed in that book in order to illustrate the rapid changes in populations even of widely distributed continental species due to changing land use and the inexorable demands on marginal wet-lands made by humans for agricultural purposes. This list of species to watch could be enlarged wherever such changing conditions occur; in South America, parts of Australia, Africa and, of course, Eurasia. In North America conditions seem to be more stable although the extensive use of pesticides, particularly in the marsh land and low lying agricultural areas of the southwestern United States, has been shown to kill ducks.

I have not listed recently extinct species or subspecies such as the pink-headed duck of India, Niceforo's pintail of Colombia, or the Auckland Island merganser. The above three forms represent examples of species which were either always rare and therefore readily disappeared from over-hunting (India), land exploitation (Colombia), or fragile island habitat and foreign predator introduction (New Zealand).

Thus all waterfowl species can be described as being threatened, essentially by man's increasing demands on the land and his perfect mobility, his persistent desire to hunt (an apparently innate irreversible characteristic), and his carelessness. At the same time waterfowl in general are highly susceptible to avicultural management and in at least six cases cited herewith have been or may be brought back from the verge of extinction by aviculturists. It is generally forgotten in America that the wood duck, *Aix sponsa,* was to a considerable extent so rescued as I described in my book *A Paddling of Ducks,* 1957 (Smithsonian reprint 1969) even though waterfowl management experts prior to World War II were oblivious to the possibilities of aviculture. The concept of keeping birds in captivity continues to be abhorrent to many bird lovers. In past years such emotions dominated the feelings of those who served as government representatives for land management for fish and game resources. Hunting was to be encouraged as a sport by officials in general but not aviculture unless it was of the kind that produced ample supplies of game birds for shooting. Nowadays there is a new, welcome sophistication in such matters and the positive values of species-saving inherent in aviculture are at last realized as the following accounts demonstrate.

Trumpeter Swan

Cygnus cygnus buccinator, Richardson

A species formerly considered greatly endangered, this our largest swan is now well established in southern Alaska, western Canada and Idaho, southwest Montana and Wyoming. Reintroduction in Nevada and South Dakota, combined with complete protection, has allowed the population to increase. Previously high concentrations of a few birds on the Red Rock Lakes and other reserves resulted in starvation or killing of young by jealously territorial adults. It is now estimated that there are between 2100 and 2500 birds of this species in the wild state as well as in captivity in North America and Europe.

Tule Goose

Answer albifrons gambelli, Hartlaub

Although this rare large subspecies of the North American white-fronted goose is known to winter in small numbers in willow-covered sloughs and small lakes in the Sacramento Valley of California, the characters of variable black color and markings on the face and the generally larger bill size and weight are still a subject of taxonomic speculation. The largest, darkest birds are those that winter near Willows, California. In 1964 Bob Elgas[1] and Jack Kiracofe discovered breeding white-fronts in willow clumps at Old Crow Flats, Yukon, in the Mackenzie River Delta. These do not appear to be identical to the Willows wintering birds, but they correspond in size and color to the type specimen of *gambelli* which Hartlaub described from "Texas and southern United States." Evidently there is a continuous cline in the population of breeding birds from far northern Canada and the Arctic slope river deltas. Typical North American white-fronts breed as far west as the Perry River north and west of Hudson's Bay, but west of that larger birds begin to be found, and there may well be a small breeding colony of the very largest, darkest form, the wintering birds of California which breed *west* of the Mackenzie. The Mackenzie River birds found by Elgas and Kiracofe may represent a population which migrates south into the country slightly to the east of the main backbone of the great chain of the Rocky Mountains.

In any case these largest and darkest North American white-fronted geese represent a very small population presumably not more than a few hundred birds. They are, of course, subject to predation from hunters as they occur throughout areas where duck and goose hunting is allowed and only an expert could distinguish the subspecies. Perhaps their preference for closed marshy ponds and lakes with fringing willows

[1] See Elgas, B. Breeding Populations of Tule White-fronted Geese in Northwestern Canada. 1970, Wilson Bull. *82* (4), p. 420–426.

(presumably in both breeding and wintering ranges) is the only thing that has saved the population. A dozen birds are in captivity.

Aleutian Canadian Goose

Branta canadensis leucopareia, Brandt

Formerly thought to be virtually extinct due to the introduction of foxes to the Aleutians, a small population of 56 birds was rediscovered on Buldir Island in 1962. From this population goslings were taken and have since been reared readily in captivity. The presence of one or two birds in flocks of migrating Canada geese in the middle flyway had prompted Carl Strutz of North Dakota to predict that a breeding population would be rediscovered, and time has proven him right. A subspecies *asiatica* Aldrich, described from Bering Island on the Siberian side of the Aleutian chain, has been determined to be inseparable from *leucopareia.*

It is understood that the American Fish and Wildlife Service will join with private collaborators in building up a sufficient population so that these hardy little surviving island Canadas may be reintroduced to other foxless islands in the Aleutian chain. This subspecies is one which formerly migrated south on either side of the North Pacific basin, reaching Japan on the one hand and California on the other. It is to be hoped that the governments of Japan and the United States will eventually extend the migratory bird treaty act convention to both countries as has been done with Canada and Mexico.

Hawaiian Goose or Nene

Branta sandvicensis, Vigors

The recovery of the Hawaiian nene from its precarious position of the mid 1940s has been phenomenal. For several years *no* wild birds were seen until Dave Woodside's sighting of a flock of 28 in 1954 off the Saddle Road on Hawaii confirmed that the species still existed in the wild. The remaining captive birds, primarily the small stock maintained by Herbert C. Shipman, has been enlarged in a spectacular manner by Peter Scott's Wildfowl Trust in England and by the Hawaiian State Game Department's farm at Pohakuloa in the bird's wild range. At present reintroductions on Hawaii from the State Game Department and on Maui Island from abroad have produced a population of close to 800 birds in the wild with another 200 in captivity, so that the species is probably on the verge of being registered as "recovered" in Red Book language, still needing to be watched and cared for, but a matchless example of concerned conservation and aviculture.[2]

Crested Shelduck

Tadorna cristata, Kuroda

The least-known species of living waterfowl, if it is living, was discovered for science only in this century by Dr. Nagamichi Kuroda, the foremost waterfowl expert of Japan. Four specimens are alleged to exist of which I have seen three, one in Copenhagen and two in Tokyo, the latter two a pair, the male of which formed Kuroda's type bird of 1917. All the records are from Korea or Vladivostock in the USSR, although old paintings found by Dr. Kuroda convinced him that this was a species and not a rare hybrid shelduck, perhaps a mating with a falcated or other species and had once occurred more widely as far as Japan. Recently in 1964 two Russian students are alleged to have sighted three birds among a flock of harlequin ducks in the Rimskii-Korsakov archipelago of islets southwest of Vladivostock (*Wildlife in Danger,* Fisher, Simon and Vincent, New York: Viking Press, 1969). The species may merely be clinging to the edge of the grave as may the nearly fabled North American Eskimo curlew.

Cereopsis Goose

Cereopsis novae-hollandiae, Latham

This species was once thought to be down to less than a thousand birds on islands south of Australia in the Bass Strait and west to the Recherche Archipelago off western Australia. In recent years more strict control of illegal shooting and some resurgence of the population on the mainland inclines Mr. Harry Frith (*Waterfowl in Australia*) to the belief that there has been a good recovery and that there may be as many as 5 to 6000 birds all told. Some 200 birds are also held in zoos and private collections. In the wild state, as in captivity, the birds are highly territorial. On islands a breeding pair has been shown to require about 1.3 acres per territory.[3]

---

[2] See "Saving the Nene. World's Rarest Goose," by S. Dillon Ripley, National Geographic Magazine, Vol. 128 (November, 1965).

[3] A recent review of the status of the species is by Dorward, D. F., The status of the Cape Barren Goose, *Cereopsis novae-hollandiae,* 1967, Xth Bulletin Internat. Council for Bird Preservation, London, pp. 56–71.

Marbled Teal

Anas angustirostris, Ménétriès

Although not listed as a threatened species in the Red Book, this little bird of Mediterranean and Middle Eastern habitat deserves careful watching. Extirpated in West Pakistan, a World Wildlife Fund project has started a reintroduction on the Lal Suhanra Reserve in Bahawalpur Division from stock bred in captivity. Changes in land use in the intervening countries seem to have greatly reduced the population of this species. It has virtually disappeared as a breeding bird from Afghanistan, from southwestern USSR, from Iraq, except in small pockets in the Tigris-Euphrates Delta and in the intervening countries west to Spain where there is a very small population in the Guadalquivir Delta. However, a wintering population of up to 2000 birds has been reported from southern Turkey in 1967–8, so that there is still an important breeding range presently unknown in Southeastern Europe or Southwest Asia. The breeding stock of captive birds derives from nine individuals imported into England from Iraq in 1948.

Madagascar Teal

Anas bernieri, Hartlaub

Found only on Madagascar, this teal has been recorded infrequently in the last half-century. A relative of the gray and chestnut-breasted teal species of the East Indies and the Australian–New Zealand Region, it is apparently restricted to and not uncommon on the west coast of Madagascar between 500 and 1500 m. elevation where three distinctive habitats, grassland savannah, mangrove, and mature forest, come together. There are no specimens in captivity.

Campbell Island Flightless Teal

Anas aucklandica nesiotis, Fleming

This subspecies of the Auckland Island duck has been questioned as being a recognizable, discrete population, but it is said to be smaller and browner than the Auckland Island subspecies. The population occurs only on Campbell Island, 500 miles south of South Island, New Zealand. A sight observation of 1958 of four birds is the last report. The subspecies is not mentioned in a recent report on birds of Campbell Island by Kinsky (1969, Notornis, Vol. 16, pp. 225–236.) The presence of introduced brown rats on the island and the reduced wings which make flight difficult (it is said to be flightless along with its close relative on the Auckland Islands) presumably have contributed to the near extinction of this form.

From the point of view of nomenclature, all four of the green-headed chestnut teal of Australia and New Zealand have been recently included as one species by the official New Zealand Checklist and by Frith (1967). These include the familiar chestnut-breasted teal of Australia, *Anas castanea,* and three New Zealand forms, *chlorotis* of the New Zealand mainland, and two allegedly flightless, *aucklandica* of the Auckland Islands and *nesiotis* the present form, on Campbell Island.

Mexican Duck
Anas diazi, Ridgway

A dull brown species, said to number perhaps 200 in the Rio Grande Valley of New Mexico, central western Texas, and southeastern Arizona. An unknown number (perhaps 1000) are said to occur in Chihuahua, Durango, and Jalisco, Mexico. The draining of marshes in the southwest United States, introduction of mallards for shooting (which have hybridized with the resident species), and, formerly, overshooting have all contributed to the decline of the species. Under the New Mexico Department of Game and Fish a captive rearing project has been underway since 1960 and a good number have been reared for subsequent release.[4]

Hawaiian Duck
Anas platyrhynchos wyvilliana, Sclater

An endangered form, currently being reared on Hawaii at the Pohakuloa Station of the State Game Department, there are also between 100 and 200 in captivity in mainland USA and Europe. The species appears to be found in the wild state today only on Kauai Island and may number fewer than 200 birds. [Recent reports of this species from the island of Oahu follow the apparent extirpation of the species on that island for the last 15 years or more. In addition, a few sightings on the main island of Hawaii up to 1959 are probably from birds released from the breeding colony at Pohakuloa. No records of breeding on either island have been obtained.]

---

[4] For a recent review of the species see Aldrich, J. W. and Baer, K. P., "Status and speciation in the Mexican Duck (*Anas diazi*)," 1970, Wilson Bull. *82* (1), pp. 63–73.

Laysan Teal

Anas laysanensis, Rothschild

Found only on Laysan Island in the Leeward Group, Hawaii, this species (or highly differentiated subspecies of the mallard) rebounded from a low of 7 birds on Laysan in 1911–12 to over 500 by 1958. Such a violent oscillation in numbers on a mile-long coral island compelled the U.S. Fish and Wildlife authorities to take cognizance of the possibility of a population crash, and in 1958 specimens were taken for captive rearing. I managed to rear young in Connecticut in 1959 as did the Wildfowl Trust in England, since which time the bird has become a common breeding species in zoos and collections. Currently there must be nearly 250 birds in captivity and the wild stock on Laysan is said to fluctuate from 400 to 600 depending on storms which sweep Laysan Island periodically. An alternative introduction of Laysan birds onto some of the neighboring islands has never been effectively attempted.

White-winged Wood Duck

Cairina scutulata, S. Müller

Very rare and decreasing in eastern Assam, India, due to forestry operations; apparently locally common in isolated teak or "sal" forest pools in Manipur, East Pakistan, and in upper Burma. A recent (December, 1970) report to the International Council for Bird Preservation from U Tun Yin of the University of Rangoon says that ducks have been observed in December in the Bhamo district of northern Burma in the Mansi and Sikaw areas. This is in addition to previous reports of 12 years ago by Oliver Milton who observed white-winged wood ducks in the Pidaung Wildlife Sanctuary in the Myitkyina district. Status unknown but said to be widely distributed (presumably very local) in Thailand and south Vietnam. May reach extreme northern peninsular Malaya. Status unknown in Sumatra and Java, Indonesia. There appear to be some 20 birds in captivity, but only one recent breeding record. Ducklings have been reared from wild eggs in Assam by Mr. M. J. S. Mackenzie under a World Wildlife Fund grant and the resulting adults distributed to the Wildfowl Trust in England and the Smithsonian Institution's National Zoo. Four additional hand-reared birds of Mackenzie's are in the Gauhati (Assam) Zoo. To be reared in captivity it appears that a large aviary is needed with preferably full-winged birds, as the species prefers to nest on platforms or stick nests in trees.

Chinese Merganser

Mergus squamatus, Gould

Although not listed in the Red Book, this local species which is said to breed in eastern coastal USSR in Ussuriland in the Sikhote-Alin region and possibly in northeast Manchuria at Chingpo Hu Lake in Kirin Province, is described as "rare, with an extremely restricted range" in Dementiev, Gladkov *et al., Birds of the Soviet Union,* Vol. 4 (English translation, 1967), p. 662. The species is unknown in captivity and presumably should be carefully watched as it may be of the same rare and local character as was the pink-headed duck of India and thus extremely vulnerable to increased hunting or changing forestry practices in its breeding habitat.

White-headed Duck or Stiff-tail

Oxyura leucocephala, Scopoli

Although not listed in the Red Book, this duck requires careful watching and its future is problematical (as pointed out by Géroudet, P., 1969, World Wildlife Yearbook, Morges, Switzerland, pp. 267–8) who estimates that less than 50 pairs remain in Europe in Spain, Corsica, the marshes of Hungary (Lake Valence), Yugoslavia, and perhaps in Rumania. A small group of less than a thousand winters in West Pakistan, particularly on lakes in the Punjab Salt Range. Isolated birds are seen from time to time wintering in northern India. Presumably these birds breed in USSR in the Sarpa Lakes area, the Volga-Ural sandy tracts east in the trans-Ural area to Kustanai and Chelyabinsk, as far as the western edge of the Altai foothills, south to the Afghanistan border marshes and the Seistan depression between Iran and Afghanistan. The latter area in Iran has been largely drained and ruined for waterfowl. Undoubtedly drainage and increasing agriculture with the use of pesticides will have a significant effect on the species. An attempt has been made to rear young from wild eggs taken in Spain and two pairs are in captivity at the Wildfowl Trust but have not bred. Stiff-tails are difficult to establish, though long-lived in captivity once accustomed to their environment. There is no essential reason why a captive population could not eventually be created even though great expertise would be required.

Waterfowl Which Have Become Extinct in the Twentieth Century

(The fact should be noted that the Giant Canada Goose, *Branta canadensis maxima* Delacour, was listed as extinct in Delacour's *Waterfowl of the World,* 1954, but has subsequently been found to be fairly common in parts of the western plains states in Minnesota, Colorado and adjacent areas.)

1. Lawrence's Brant Goose, *Branta bernicla nigricans* Lawrence. Bred in northeast Hudson's Bay? Wintered off New Jersey. Thought to be extinct.

2. Niceforo's Pintail, *Anas georgica niceforoi* Wetmore and Borrero. Highlands of Colombia from Bogota south. Thought by Borrero to be now extinct from over-shooting and habitat destruction.

3. Coues' Gadwall, *Anas strepera couesi* Streets. Washington Island south of Hawaii. Known only from the original two specimens taken in the 1870s.

4. Borrero's Cinammon Teal, *Anas cyanoptera borreroi* Snyder and Lumsden. Highlands of Colombia. Now thought by Borrero to be extinct from over-shooting and habitat destruction.

5. Pink-headed Duck, *Rhodonessa caryophyllacea* Latham. Gangetic drainage basin of eastern India and East Pakistan. Extinct since 1939 (captivity); last record in the wild in Bihar in early 1930s.

JANET KEAR

Hawaiian Goose or Nene

The Hawaiian goose or nene is one of three members of the Anatidae restricted to the Hawaiian archipelago. All three are currently listed in the International Union for the Conservation of Nature Red Book as being rare and endangered. Two are ducks that have features peculiar to remote island forms but are obviously still dabbling waterfowl. The Hawaiian goose, on the other hand, has become highly modified by its life in open, sparsely vegetated, volcanic areas, where it finds standing water only in temporary rain pools.

In the 18th century, the Hawaiian goose, commonly called the nene, was widespread below 9000 feet on the tropical islands of Maui and Hawaii, living between latitudes 19°N and 21°N. By the 1950s, its numbers had dropped from 25,000 to under 50, and only an estimated 35 individuals remained in the wild. A variety of man-made factors probably contributed to that dramatic decline: like many island creatures, the bird is potentially tame and after the introduction of firearms it was easily killed. The breeding season begins as the hours of daylight grow less, an unusual feature presumably based on the period at which the best food supply is available for the goslings. In Hawaii, the first eggs are laid in late October and an extended breeding season may last into February. One reason for the species' decline was a winter open season for hunting, coinciding with the birds' nesting activities. The white man shot during autumn and winter in his temperate home and expected tropical game birds to fit the pattern of springtime egg-laying to which he was accustomed. A further upset was that the lowland breeding areas were drastically altered by the encroachment of sugarcane plantations and the grazing of sheep and goats. Pairs nesting in the higher lava fields suffered from the attacks of newly introduced predators such as the mongoose, rat, wild pig, cat, and dog. Even today, when the nene is rigidly protected—since 1957 it has been Hawaii's state bird—exotic predators are still a threat to its survival. The mongoose, for one, will probably never be eliminated, and without strict and constant control of predators at the nesting sites, the wild goose population remains at risk.

The history of nene conservation has become something of a classic. A combination of good fortune, the farsightedness of certain Hawaiian agencies, and the fact that the species will readily breed in captivity has contributed to its success. As long ago as 1902, H. W. Henshaw pointed out the foolishness of the winter shooting season and the need for planning to save the nene from extinction. It was not until 50 years later that an official survey was sponsored by the Hawaiian Board of Agriculture and Forestry. The objectives were to study the biology of the wild population, to prepare management recommendations, and to assess the efforts being made by the Board in Hawaii and the Wildfowl Trust in England to rear nene geese in captivity. The Board began breeding in 1949 with two pairs of geese, and a third pair was added in the following year. The Wildfowl Trust birds, two females and a male, bred for the first time in 1952. All these geese were lent to the breeding projects by Mr. Herbert Shipman, a Hawaiian ranch owner who, realizing the perilous state of the species, had maintained a small flock of nene geese on his property.

Initially, the production of goslings in captivity was rather disappointing. At both breeding centers, the first and second clutches of eggs were removed and hatched by foster mothers (hens or muscovy ducks). Some of the geese then laid repeat clutches. Third clutches are not uncommon in England, and occasionally four clutches a season are obtained in Hawaii. Despite the high fecundity of the females, a large number of the eggs laid in the 1950s failed to hatch, apparently because they were infertile. However, by 1955 there were probably more birds in captivity than in the wild, and nearly one-third of the world's population was in England.

In 1959, the federal authorities of the United States of America awarded a grant to the Nene Restoration Project, and 1960 saw the first release of captive-reared geese to augment the wild flock. An area had been acquired near the known breeding ground and surrounded by a predator-proof fence. Birds for release were put into the fenced area with clipped wing feathers, so that they could become conditioned to natural foods while also receiving their routine diet. After the molt and the growth of new feathers, they gradually flew farther and farther from the release pen, at first returning to it at night. Finally they joined the wild birds during their normal flocking season.

The problem of low fertility continued to plague the captive-rearing program. It seemed likely that this had arisen from inbreeding, since all the captive stock in the world stemmed from a few original pairs in Mr. Shipman's flock. It was felt that an introduction of new stock was essential, and a few nene eggs were collected in the wild and hatched at the Pohakuloa headquarters of the restoration project. The goslings obtained were later paired with captive geese both in Hawaii and England. This out-breeding has had considerable success and fertility levels have risen at both centers.

Improved incubation procedures, sound nutrition, and vigorous selective breeding have also helped. Males known to be fertile, or males with particularly fertile fathers, were paired with females known to lay a large number of eggs each season, or with their daughters. Relatively unsuccessful nests and their offspring were eliminated from the breeding program. As a result, the fertility of eggs laid at the Wildfowl Trust rose from around 40 percent to over 80 percent in four years. Interestingly, the average weight of eggs laid also increased. Since heavier eggs were found to have a better chance of hatching, it was apparent that the performance of the females had been improved alongside that of the males.

The production of goslings increased dramatically. The first release of thirty English-bred nenes was made in 1962, not on the island of Hawaii, but on Maui, where the species had become extinct during the 19th century. That was done to ensure that the remaining wild birds would not be contaminated by any undesirable features acquired by the English stock. The geese were, of course, as disease-free as careful treatment and long quarantine could make them. Releases have continued on both islands; by the summer of 1968, 336 Pohakuloa-reared birds had been released on Hawaii and 170 on Maui, the majority of them English.

In terms of breeding success, birds that are able to join a wild flock appear to be at an advantage, probably because traditional feeding patterns, flight lines, and nesting areas are thus readily acquired. However, nests of nene geese released after ringing have been found on both islands and increasing numbers of juvenile geese are being seen. In Hawaii, wild-bred birds are mating with captive-bred ones and rearing young successfully together.

The restoration project is continuing—50 nene left England in July 1969 and 56 more in 1970. It seems likely that a real success has been achieved in the augmentation of a vanishing wild species from captive sources. In any case the nene goose has been saved from extinction; the world population is now over 900 birds and the current annual production of goslings in captivity is far in excess of adult mortality. One hundred goslings were produced in Britain in 1970. A direct achievement has been the arousal of tremendous public interest in the nene and, thereby, an increased awareness of the value and possibilities of other conservation programs. To that end, pairs of nene geese have been distributed to about 35 zoological gardens and wildlife collections in Europe and North America. As well as making attractive and educational exhibits, the birds help to ensure that no single disaster overtakes the whole population in future.

Of course, problems still exist and further research is being undertaken into the fertility and hatchability of eggs and into the control of breeding conditions and good health in the adult birds. Much of that research will also be of great value to other restoration projects.

List of Contributors

BAUER, ROBERT V.—Formerly manager of waterfowl propagation for New York State Conservation Department. He now owns a game bird farm in Lapeer, Michigan.

BLOCK, MELVIN L.—For 35 years Curator of Birds at the Toledo Zoological Gardens, Toledo, Ohio.

BRUNING, DONALD—Assistant Curator of Ornithology at the New York Zoological Park.

CECIL, MARVIN—Former Manager of Caribbean Gardens, Naples, Florida. Under Cecil's management, the waterfowl collection grew to be the second largest in the world, numbering 148 species and subspecies of Anatidae. First Chairman of the International Wild Waterfowl Association Book Committee.

CONWAY, WILLIAM G.—General Director of New York Zoological Society and Curator of Birds of the New York Zoological Park.

DELACOUR, JEAN THEODOR—Director Emeritus, Los Angeles County Museum. Research Associate, American Museum, New York and Smithsonian Institution, Washington. Collaborator U.S. Fish and Wildlife Service. French National Museum. President of the International Wild Waterfowl Association. Co-author with Peter Scott of *Waterfowl of the World*.

ELGAS, ROBERT—Rancher, artist, aviculturist from Big Timber, Montana. Authority on Tule Geese.

ELLIS, JOHN H.—Aviculturist from Hartford, Connecticut. Mr. Ellis has reared several species for the first time in North America and three species for the first time in the world.

FERGUSON, WILLIAM H.—Ophthalmologist. Compiler of this book. Dr. Ferguson's articles on aviculture have appeared in many avicultural publications.

FLIEG, G. MICHAEL—A roving curator of birds and collector in the field whose assignments have included the St. Louis and Chicago (Brookfield) Zoos and the private collection of Mr. Michael Dam, Haines City, Florida.

FOLEY, DON—Biologist at Patuxent Wildlife Research Center, Patuxent, Maryland.

FRITH, H. J.—Chief of the Division of Wildlife Research, CSIRO, Canberra, Australia. Author of *Waterfowl in Australia.* Awarded Doctor of Agricultural Science, by University of Sydney in 1964 for work on bird ecology.

GENTRY, CYRILLE—Klamath Falls, Oregon, artist, whose pen-and-inks and oils of western, Indian and wildlife subjects have been featured in several books, magazines, museums, and galleries.

GRANT, ROBERT W.—Aviculturist of Laurel, Maryland. Former Director and President American Pheasant and Waterfowl Society.

GREENWELL, GUY A.—Breeder of waterfowl since 1925. Former biologist with Missouri Department of Conservation. Now Curator of Birds, National Zoological Park, The Smithsonian Institution, Washington, D.C.

GRISWOLD, JOHN A.—Curator of Birds, Philadelphia Zoological Gardens, 1947 to present. Former Research Assistant in Ornithology, Museum of Comparative Zoology, Harvard. Scientific expeditions to Yugoslavia, Panama, Thailand, Borneo, Peru, Yucatan, and Ecuador. Authority on cranes and waterfowl.

GUEST, WINSTON—Owner of Duck Puddle Farm, Oyster Bay, Long Island, New York, currently the largest private collection of waterfowl in North America.

HEFT, ELMER—A veteran of thirty-three years of raising waterfowl, he maintains a large collection in the Middle West.

HOCHBAUM, H. ALBERT—Director, Delta Duck Station, Delta, Manitoba, Canada.

HOLLENBECK, WILLIAM—A long-time aviculturist from Parma, Ohio, specializing in swans.

HOLM, EARL—Supervising Wildlife Biologist New York State Department of Environmental Conservation. Mr. Holm has long been noted for his knowledge of waterfowl.

HUEY, WILLIAM—Mr. Huey is currently Assistant Director of the New Mexico Department of Game and Fish. A specialist in waterfowl research and habitat development, he is an authority on the New Mexican duck and was the first to propagate this species.

HYDE, DAYTON O.—Aviculturist, editor, author of *Sandy, the True Story of a Rare Sandhill Crane Who Joined Our Family; Yamsi; Cranes in My Corral.* Conservationist and Oregon cattle rancher.

JOHNSGARD, PAUL A.—Professor of Zoology at the University of Nebraska. Dr. Johnsgard, an eminent authority on waterfowl, is the author of four books, including *Handbook of Waterfowl Behavior* and *Waterfowl: Their Biology and Natural History.*

KEAR, JANET—Aviculturist at the Wildfowl Trust, Slimbridge, England. Dr. Kear is an eminent authority.

LEE, FOREST—Chief of Section of Techniques Development at Northern Prairie Wildlife Research Center, Jamestown, North Dakota and in charge of waterfowl propagation at N.P.W.R.C.

LINT, KENTON C.—Since 1948, Mr. Lint has been Curator of Birds, Zoological Society of San Diego, California. His special interest is the field of conservation and propagation of birds in captivity. His record includes forty-six species of waterfowl, four hundred breeding records. He is the President of the Whooping Crane Conservation Association and a director of many others. His expeditions have taken him to Australia, Indonesia, Borneo, Thailand, and Central and South America.

LUBBOCK, MICHAEL—Curator of Birds at the Wildfowl Trust, Slimbridge, England. From 1969 to 1972 Mike Lubbock was on loan as Curator of Birds for Mr. Winston Guest, Duck Puddle Farm, Oyster Bay, New York, where he achieved an enviable record in the breeding and care of waterfowl in this extensive collection. He has participated in collecting expeditions to all parts of the world.

MACY, WILLIAM K.—As a boy, Mr. Macy collected native waterfowl. After a career as a broadcasting executive, he has rebuilt his collection on Long Island, New York, and raises a good many of the rarer species.

MILLER, GEORGE W.—Mr. Miller is a leading aviculturist from Ayr, Ontario, Canada. His work with waterfowl has won him many breeding awards.

MULLER, KERRY—Curator of Birds, Taronga Park Zoo, Sydney, Australia. Formerly at the San Diego Zoo, then Manager of Bird Department at National Zoological Park, Washington, D.C.

RIPLEY, S. DILLON—Mr. Ripley is Secretary of the Smithsonian Institution, Washington, D.C. He has been rearing waterfowl since 1931, and has written twenty-three papers on waterfowl. Currently Mr. Ripley is also President of The International Council for Bird Preservation.

ROOTS, CLIVE—Aviculturist at Winged World, Heysham, Lancaster, England.

STRUTZ, CARL—Administrative Officer at the Northern Prairie Wildlife Research Center, in Jamestown, North Dakota. Currently Secretary-Treasurer of the International Wild Waterfowl Association, he has raised Giant Canada geese for over twenty years, taking over a flock his father started in 1893.

SCHEDEL, J. J.—Doctor of Public Service, Bowling Green State University, Bowling Green, Ohio. Dr. Schedel has long been a friend of waterfowl research. A board member of the Toledo Zoological Society, he started his collection in 1957 in Elmore, Ohio, and is one of the very few to have raised four generations of Australian shelducks and two generations of Coscoroba swans in the Northern Hemisphere.

SCOTT, M. L.—Professor of Nutrition and Poultry Husbandry, Cornell University, Ithaca, New York.

SEARLES, GEORGE—A director and officer in several avicultural organizations, Mr. Searles maintains an extensive waterfowl and game-bird collection in Patchogue, New York, and is a successful breeder of several rare species including the harlequin duck.

SOUTHWICK, JUSTIN A.—A long-time member of avicultural organizations, Mr. Southwick was the owner of the Southwick Animal Farm in Blackstone, Massachusetts. A great lover of birds and animals he was born and died on the same homestead. At the time of his death in 1971, he was a director of the International Wild Waterfowl Association.

SKINNER, JOHN L.—Extension Poultryman, University of Wisconsin at Madison.

SMART, GLEN—Biologist and aviculturist in charge of rearing endangered species at the U.S. Bureau Sports Fisheries and Wildlife, Wildlife Research Center at Patuxent, Maryland. Mr. Smart is noted for his waterfowl carvings.

STANTON, PHILIP—Assistant Professor of Biology at Framingham State College, Framingham, Massachusetts. Mr. Stanton breeds waterfowl and is Director of the Wildlife Rehabilitation Center, a waterfowl research facility specializing in care and rehabilitation of oil-soaked sea birds.

WALLACH, JOEL D., D.V.M.—Associate Director at the Chicago Zoological Park, Brookfield, Illinois, and formerly Director of Research at the St. Louis Zoological Gardens. He is father and editor of *The Journal of Zoo Animal Medicine.*

WELLER, MILTON W.—Dr. Weller is Professor of Zoology and Chairman of the Fish and Wildlife Biology Section of the Department of Zoology and Entomology, Iowa State University, Ames, Iowa, and has done research on waterfowl in Missouri, Iowa, Manitoba, Arctic, Antarctic, Falkland Islands, and South Georgia Islands.

WELLS, DONALD P., JR.—Breeder of North American waterfowl and currently engaged in research, manufacture, and marketing of water-pollution control equipment in Lake Bluff, Illinois.

WEBSTER, CHARLES D.—Chairman of Book Committee, a waterfowl breeder on Long Island with many rare species, and President of the Horticultural Society of New York.

WILSON, CALVIN—For many years Mr. Wilson has been manager of the Tracy Aviaries in Salt Lake City, Utah. His articles have appeared in many aviculture magazines.

Waterfowl Publications

Although the following references are regarded as especially pertinent for supplementary reading, numerous additional references can be found in the fourth volume of Delacour's *Waterfowl of the World* (over 900 references up to about 1962), in Phillips' *Natural History of the Ducks* (over 2700 references through 1924), and in N. Kuroda's *Bibliography of the Duck Tribe* (over 6500 references through 1940). The U.S. Fish and Wildlife Service's *Wildlife Review* has abstracted over 1900 references on waterfowl from 1935 through 1966. The *Aves* section of *Zoological Record* (Zoological Society of London) annually indexes most English-language publications that concern the Anatidae, and the most recently published volume (covering 1964 literature) has more than 200 references dealing wholly or in part with species of this family.

Boyd, Hugh, and Scott, Peter (eds.). *Annual Reports of the Wildfowl Trust.* 18 vols. Slimbridge, England: The Wildfowl Trust, 1950–67.

Delacour, Jean. *The Waterfowl of the World.* 4 vols. London: Country Life, 1954–64.

Johnsgard, P. A. *Handbook of Waterfowl Behavior.* Ithaca, N.Y.: Cornell University Press, 1965.

Kortright, F. H. *The Ducks, Geese, and Swans of North America.* Washington, D.C.: American Wildlife Institute, 1943.

Phillips, J. C. *The Natural History of the Ducks.* 4 vols. Boston: Houghton Mifflin, 1922–1926.

Wetmore, Alexander (ed.). *Water, Prey, and Game Birds of North America.* Washington, D.C.: National Geographic Society, 1965.

Hochbaum, H. A. *Travels and Traditions of Waterfowl.* Minneapolis: University of Minnesota Press, 1955.

Lorenz, K. Z. (1951–1953), "Comparative Studies on the Behaviour of the Anatinae," *Avicultural Magazine,* 57:157–182; 58:8–17, 61–72, 86–94, 172–184; 59:24–34, 80–91.

Frith, H. J. "Ecology of Wild Ducks in Inland Australia," in *Biogeography and Ecology in Australia.* Den Haag: W. Junk, 1959.

Greenway, J. C., Jr. *Extinct and Vanishing Birds of the World.* Special Publication No. 13, American Committee for International Wildlife Protection, 1958.

Scott, Peter. *A Coloured Key to the Wildfowl of the World.* Slimbridge, England: The Wildfowl Trust, 1957.

Sowls, L. K. *Prairie Ducks.* Washington, D.C.: Wildlife Management Institute, 1955.

Hochbaum, H. A. *The Canvasback on a Prairie Marsh,* 2d ed. Washington, D.C.: Wildlife Management Institute, 1959.

Index

Humidity pg. 96